THE

NEW ASTRONOMY GUIDE

STARGAZING IN THE DIGITAL AGE

PATRICK MOORE AND PETE LAWRENCE

FOREWORD BY
BRIAN MAY

CARLTON
BOOKS

THIS IS A CARLTON BOOK

This edition published in 2015 by Carlton Books Limited

First published in 2012 by Carlton Books Limited
20 Mortimer Street
London W1T 3JW

10 9 8 7 6 5 4 3 2 1

A CIP catalogue record for this book is available from the British Library.

ISBN 978 1 78097 613 6

Printed in China

Created for Carlton Books by Canopus Publishing Limited
15 Nelson Parade
Bristol, BS8 4HY
UK
www.canopusbooks.com

For Canopus Publishing Limited:
Director and Editor: Robin Rees
Additional Editorial: Tom Jackson
Designer: Jamie Symonds
Art: Jamie Symonds
Photographs: (unless otherwise specified) Pete Lawrence
Star charts: Pete Lawrence
Author photographs: Jamie Symonds

For Carlton Books:
Executive Editor: Gemma Maclagan Ram
Additional Editorial: Barry Goodman, Catherine Rubinstein
Design Manager: Russell Knowles
Design: Russell Knowles, Kate Painter
Production: Maria Petalidou

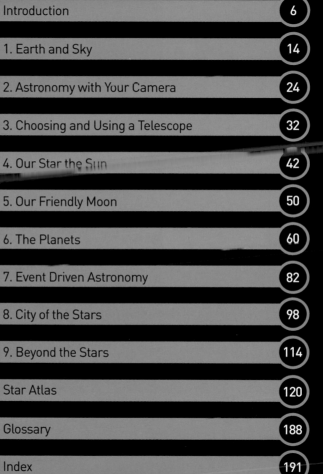

PREFACE

Astronomy is one of the oldest of the sciences, and even today most of the names we use for stars date back to the time of the ancient civilizations. The recent surge in media coverage has helped create an explosion of interest in the subject, and thanks to modern technology, there is no longer a daunting boundary to cross before you can begin to do astronomy for yourself.

What has changed beyond recognition in recent years is the use of readily accessible technology in astronomy - from the digital camera, to computer control for telescopes, from the Charge Coupled Device to the vast array of software available to assist astronomers to process their images.

Our motive in writing this book is to combine or the first time the science and practice of astronomy, with an account of the latest technology and techniques. Advice is given on chosing and using a telescope, teaming it with a camera, and capturing the sort of images which until recently could only be obtained with professional instruments.

At the back of the book is a full star atlas to help you track down what you can see on any night of the year, making this a complete and self-contained guide book.

Patrick Moore
Pete Lawrence
Selsey, UK, 2012

FOREWORD

This beautiful book is the result of a perfect collaboration – bringing together two of Britain's most accomplished astronomers.

To anyone brought up in the UK in the last 60 years, Sir Patrick Moore - the father of modern astronomy in Britain - needs no introduction. I cannot think of an astronomer, professional or amateur, who has not, in some private moment, confided that Sir Patrick was a prime inspiration to them to enter a life-long study of the Universe. Patrick Moore's timeless BBC TV programme, The Sky at Night, is still breaking records after 55 continuous years of broadcasting; it continues to communicate all that is new and challenging in astronomy and astrophysics, and stands as a monument to Patrick's devotion to the subject. In those 55 years, astronomy has changed and broadened immeasurably, and these days it has become almost impossible for any one scientist to be in touch with developments in all the branches of current astronomical research. Yet Patrick has managed to do just this. Try him with a question about any part of the Universe, and he will answer not with a reference to some source in the Internet ... but with an account which would make you swear that he has been there, and seen it with his own eyes. It is this intimacy with the Universe that still sets Sir Patrick Moore apart from those who follow him ... and it is this precious quality which you will find embedded in The New Astronomy Guide – Stargazing in the Digital Age.

Pete Lawrence, in a career in astronomy that has already spanned 20 years at the top of his profession, has become a regular presenter on Sky at Night, and leading exponent of astrophotography. His mastery of the technical side of astronomy and photographing the Heavens is matched by an unerring instinct for capturing the awesome beauty of Universe around us. Pete's dual expertise in computing has enabled him to bring out the best in the new digital technology that has swept through astronomical imaging, and he has developed a deep understanding of every advance in digital imaging along the way.

These two experts are the perfect foil for each other; they have been close friends and colleagues for many years, but here for the very first time they have worked together to create a unique portrait of modern astronomy through astronomical photography - its equipment, its methods, and some of the medium's most spectacular results.

We humans have the gift of very fine eyesight, and from the earliest times our ancestors - nomads, or shepherds on distant hills under dark skies - have seen clues in the night sky as to what is out there. But what our eyes cannot do is add up light signals over a period of time. No matter how long we stare at a distant galaxy, it will never get any brighter, and we will never be able to see it more resolved than a fuzzy patch in our field of vision. In fact, as every observational astronomer knows, we actually get a slightly better view if, rather than staring, we avert our vision slightly. But part of the inherent magic of photography, from its very birth in the hands of Niepce, Daguerre and Fox-Talbot in the 19th century, is its ability to accumulate the effect of light. The longer a sensitive film is exposed to light, the brighter the image gets. And modern digital arrays of light-sensitive elements have been designed to reproduce that ability - to integrate light over minutes, hours, and even weeks, building up an image of ever-increasing depth and detail. And so the fact that we all now know what the beautiful spiral of the Andromeda Galaxy actually looks like is not due to anyone's vision, looking through a telescope. Every one of those images we now have in our heads is due to long-exposure photography. Of course, we Earthlings are observing from a spinning ball, so nothing in the sky stays still, and keeping an image in one place to achieve these long exposures without blurring is one of the problems astrophotographers face; but there are other obstacles in the way ... not least the turbulent atmosphere above us - which makes the stars twinkle to the naked eye, and jump about alarmingly in a telescope.

These and many other matters are discussed in this book, which provides advice and information in abundance on every page. Along the way, the reader will delight in Sir Patrick's up-to-the minute information on each astronomical subject, while enjoying it in its full glory through the photography of Pete Lawrence and other leading exponents of the art.

Enjoy - and marvel, all ye who are fortunate enough to enter here!

Brian May, 2012

This carefully timed shot uses clouds to dim the
brightness of the Moon, allowing its disc to be properly
exposed while also capturing Jupiter and its moons.

Our astronomical exploration of the Universe starts in the heart of our Solar System with our own star, the Sun. From here we introduce the fascinating and beautiful astronomical bodies that you can enjoy with the naked eye: the Moon, the planets, comets and meteors, and the myriad of stars within our own Milky Way Galaxy that are arranged into the familiar patterns of the constellations.

It may come as a shock to some to discover that our Sun, a huge ball of incandescent gas which could contain a million bodies the size of the Earth, is a normal star. The surface is at a temperature of 5–6000 Celsius, and near the centre the temperature rises to millions of degrees at immense pressure. The Sun is not "burning" in the ordinary sense of the word. The most plentiful substance in the universe is hydrogen, lightest of gases, and the Sun contains a great deal of it. Near the core, where the temperatures and pressures are so high, strange things are happening. Atoms of hydrogen (H) are combining to form atoms of helium (He), the second-lightest gas. It takes four atoms of H to make two atoms of He, and each time this happens a tiny amount of mass is lost, being converted to energy. This energy keeps the Sun shining, and the loss of mass amounts to more than 4 million tonnes per second, so the Sun weighs far less now than when you opened this book. Don't panic! Enough hydrogen is left to maintain the Sun for thousands of millions of years. We say more about the Sun in Chapter 4.

All the stars we see in the night sky are themselves "suns", many of them much more luminous than our own Sun, but, of course, they are a great deal further away. Here we must introduce a unit we will use throughout this book – the light year. We know that the Sun is 93 million miles away from the Earth, and for the distances inside our Solar System, our everyday units such as the mile and kilometre are suitable; in deep space they are not. Even the nearest star beyond the Sun is over 24 million million miles away. Trying to express distance of this sort in miles or kilometres would be as clumsy as giving the distance between London and New York

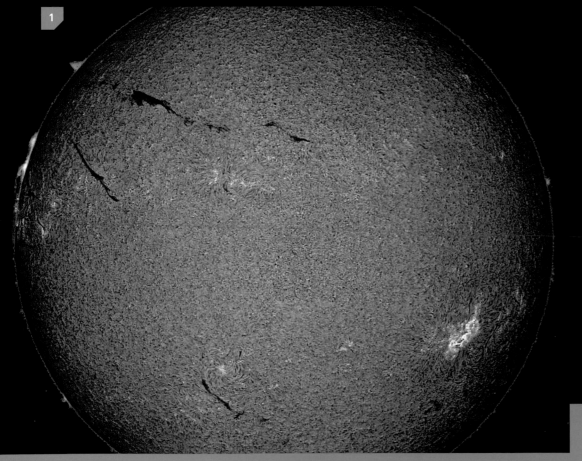

[1] An image of the Sun taken through a hydrogen-alpha filter.

[2] The planets Venus and Jupiter close together in a winter evening sky.

[3] A thin crescent Moon showing earthshine and the planet Mercury lie close to the Pleiades open cluster.'

[4] Detail of the planet Mars captured with a 14-inch amateur telescope.

in inches. Luckily, nature has provided us with a much better unit. Light does not travel instantaneously, but flashes along at the staggering rate of 186 thousand miles per second.

This means that in one year, light can travel almost 6 million million miles, and this distance we call a light year – which, please note, is a measure of distance, and not of time.

Most of the stars you see at night with the naked eye are many light years away. For example, one bright star prominent in the winter time, named Rigel, is over 700 light years away, and more than 40 thousand times more powerful than our Sun. At that distance it shrinks into a tiny point. A further phenomenon here: stars appear to twinkle. This is solely due to the effects of Earth's atmosphere. Seen from space or the surface of the Moon, the stars no longer twinkle.

Introducing the Solar System

The Sun is at the centre of the Solar System, of which the main bodies, apart from the Sun itself, are the eight planets. Our Earth is an ordinary planet, third in order of distance from the Sun. Two planets, Mercury and Venus, are closer to the Sun than we are, and the rest, Mars, Jupiter, Saturn, Uranus and Neptune, are much further away. Superficially, planets look like stars, and some are very brilliant – particularly Venus. It has been said that while stars twinkle, planets do not. A star appears as a point of light while a planet is a disc. All the same, when a planet is low in the sky and its light is coming through a thick layer of our atmosphere there is a definite twinkling effect.

Apart from Mercury and Venus, all the planets are tended by secondary bodies known as satellites. Earth has one satellite, our familiar Moon, but other planets have more; Jupiter, the largest planet, has four main moons and a whole host of smaller ones. Also in the Solar System we find various minor bodies, notably the asteroids. They are small, usually less than 100 miles across, and most orbit the Sun between the paths of Mars and Jupiter. Right at the edge of the main Solar System, beyond the planet Neptune, we have another ring of small bodies, known as the Kuiper Belt, in honour of the Dutch astronomer Gerard Kuiper, one of the first to recognize their existence and to study it.

Comets are quite different from planets; they move round the Sun, but most do so in long, highly eccentric orbits so their periods may amount to many centuries. A typical comet consists of a nucleus of ice and rocky particles that becomes surrounded by layers of gas and dust as the nucleus warms. There are many periodical

comets that orbit the Sun in periods of a few years, but really bright comets are rare visitors. Some have become so bright they even cast shadows. Note, of course, that planets and comets shine only by reflected sunlight; they have no light that they themselves generate.

Most people have seen shooting stars or meteors. A meteor appears as a streak in the sky that may last for a few seconds before disappearing. It has absolutely nothing to do with a real star. A meteor occurs when a tiny particle known as a meteoroid, typically a few millimetres to centimetres in size, dashes into the Earth's atmosphere and vaporizes, leaving only fine dust. Some meteoroids collect in swarms, and every time we pass through such a swarm, meteors are plentiful. Now and then the Earth is hit by a much more massive body which lands intact and is then known as a meteorite. There is no connection between a meteor and meteorite. Meteors are cometary debris, but meteorites are made up of stone, iron or both. In the past there have been really major impacts. About 65 million years ago it is widely believed a huge meteorite landed in the Yucatan Peninsula and caused such devastation that the dinosaurs died out in the resulting conditions. This theory is by no means proven, but it is well supported and plausible. What has happened in the past can happen again, and we will be hit by a major meteorite. We hope it will not be in a densely populated area. If a meteorite say half a mile across landed in the middle of London the death toll would be catastrophic. Fortunately the chance is slight.

[5] A Geminid meteor streaks across the December night sky.

[6] The magnificent constellation of Orion, the Hunter, rising in the east.

[7] Comet C/2006 P1 McNaught photographed from the UK in January 2007.

[8] A long-exposure photograph of the Plough asterism results in star trails.

[9] The beautiful double star Albireo in the constellation Cygnus.

On to the Stars

Beyond Neptune and the Kuiper Belt, it's believed there's a huge region of cometary objects known as the Oort Cloud. Then there's a vast gap before we come to the nearest star. The first star maps were drawn up many centuries ago, notably by the Greeks, and the stars were divided into "constellations", many of which had mythological names. One of the best known constellations is Ursa Major, the Great Bear. Astronomically we use the Latin names, for Latin is still the universal language even though no one speaks it. The seven main stars of Ursa Major are spread out in the pattern we often call the Plough or (in America) the Big Dipper.

All the main stars are many tens of light years away and, although they are moving through space at high speeds, their individual or proper motions are so small they can only be noticed over periods of many lifetimes.

Come back to the Earth in 10 million years time, and the pattern of the Great Bear will be distorted because two of the main stars are moving in the opposite direction to the other five. Note that a constellation is purely a named pattern of stars and has no real significance because the stars are at very different distances away from us.

The constellation Orion, the Hunter, dominates the sky in winter evenings in the northern hemisphere, and the two main stars, known as Betelgeux and Rigel, are particularly brilliant. They are placed in the same constellation but there is no connection between the two. Rigel is much the more remote. From another vantage point many light years from Earth, they may appear to lie on opposite sides of the sky. The main constellations were drawn up in 1651 and were given mainly mythological names. Other constellations have been added since and given modern names, such as Microscopium (the Microscope) and Telescopium (the Telescope). This is particularly true for the stars in the far south of the sky that couldn't be seen from Europe when the first maps were made.

Earth spins on its axis once in 24 hours and does so from west to east. This means the entire sky seems to move round from east to west carrying the Sun, Moon and planets with it. In a northward direction the axis points to the north celestial pole, closely marked by Polaris in the constellation of Ursa Minor, the Little Bear. This means that Polaris stands almost stationary while everything appears to move round it every 24 hours. South of the equator Polaris can't be seen. The southern equivalent, Sigma Octantis, is further from the south celestial pole, and is much fainter.

There are stars of many kinds. Most shine constantly for year after year, century after century, but some brighten or fade over periods of days, weeks or months. These are the variable stars. Stars very often appear in pairs, with components that may be equal or very unequal in brightness. These double stars are of two kinds: optical doubles are line-of-sight effects, with one more-or-less behind the other. However, in most cases the components are connected and move round their common centre of gravity. These are the binary stars. Rather surprisingly they are more common than optical doubles. Our solitary Sun is exceptional.

The Milky Way

All the stars we can see at night with the naked eye are members of our own galaxy. Our Galaxy is a flattened star system containing 100,000 million stars. When we look along the plane of the Galaxy we see many stars in almost the same direction and these form the band in the sky known as the Milky Way. The stars of the Milky Way are so numerous

they appear almost to be touching. Appearances are deceptive, as so often in astronomy, and on average the stars are light years apart.

We also see clusters of stars, and here again they are of two kinds – some are pure line-of-sight effects while others are genuinely associated. The best known example is that of the Seven Sisters or Pleiades, prominent in northern hemisphere winter skies, in the evening. People with average eyesight can see at least seven stars, while those with keen sight can see more – the record is 19. The whole cluster contains several hundred stars, which were created in the same area at the same time. Rather different are the globular clusters, huge spherical systems that may contain up to a million stars. They are so far away few are visible with the naked eye. The best example in the northern hemisphere at the latitude of the UK or the northern United States is Messier 13, M13, the Hercules Cluster.

[10] The globular cluster Messier 13 in Hercules.

[11] The Pleiades open cluster (Messier 45), taken by Ian Sharp.

[12] The Markarian Chain, part of the Virgo Galaxy Cluster, taken by Ian Sharp.

[13] Nebulosity within Orion's Sword including the impressive Orion Nebula (Messier 42).

Cataloguing the Universe

In 1781, the French astronomer Charles Messier drew up a catalogue of over 100 star clusters and nebulae. He did this not because he was interested in the objects themselves, but rather because he wanted to avoid them. He was a comet hunter and it is easy to confuse a faint nebula or cluster with a comet. In modern times, Messier's catalogue is still used to list out some of the brightest and best deep-sky objects in the sky. However, from an astronomical point of view, the catalogue is woefully short and the myriad of such objects now known to exist is better covered by the far more extensive New General Catalogue (NGC) and Index Catalogue (IC) lists.

As an aside, PM was once in his observatory observing Jupiter, and when Jupiter was obscured by a cloud, amused himself by looking at some of the clusters and nebulae not listed by Messier, either because they could not possibly be

confused with comets or because they were too far south in the sky to be seen from France. Light-heartedly he compiled a catalogue of more than 100 objects and called it the Caldwell catalogue. Caldwell being part of his surname – a hyphenated one, Caldwell-Moore. To his surprise, these "C" numbers are now widely used.

If we observe nebulae, which is Latin for "clouds", through a small telescope, they look like patches of light in the sky, and are of two main types. Some, such as the Great Nebula in the constellation of Orion, look as if they are made of gas, and in fact they are. In nebulae of this kind, new stars are being formed from the interstellar material. Other nebular objects, such as M31 in the constellation of Andromeda, are galaxies in their own right containing millions of stars.

Our Galaxy has about 100,000 million stars, and from above would look like a spiral. Beyond our Galaxy, many millions of

light years away, we see other galaxies, some spiral, some elliptical and some irregular. Each contains its quota of stars, and the total number of stars in the universe is staggeringly great. Moreover we know that many of these stars are attended by planets in the same way as our own Sun. This means there must be millions upon millions of planets like Earth, and who can tell if they support life? This again is a subject to which we will return later. It is very hard to believe we are unique.

Also, we cannot see right to the edge of our universe – assuming it has an edge, which is far from certain. The most distant objects we can see are 13.7 thousand million light years away. What happens even further out, we do not yet know.

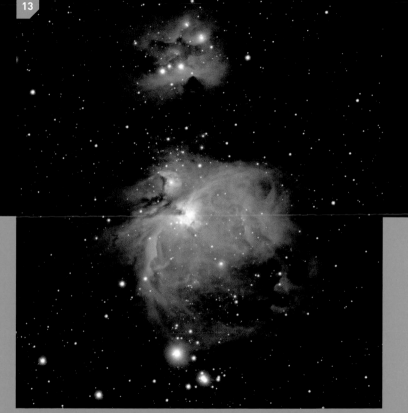

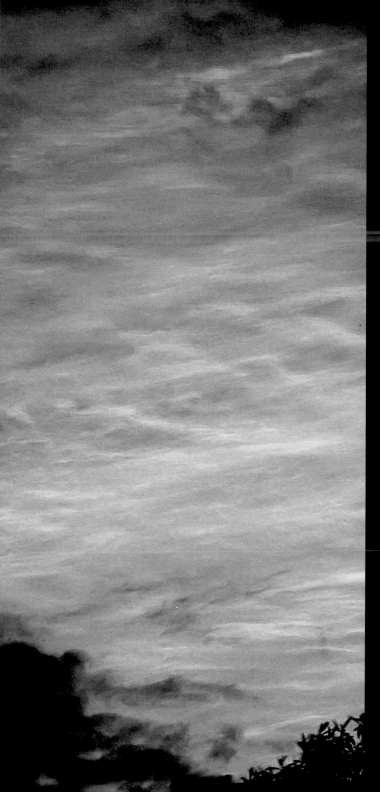

EARTH AND
SKY

Noctilucent clouds.

CHAPTER 1

In this chapter we look at a diverse range of phenomena involving light and particles. Some of these phenomena are purely astronomical, while others are meteorological, due to atmospheric effects, and some cross the boundaries between the two. What they have in common is an extraordinary beauty, making them very desirable both to look at and to photograph.

Red and Blue Sky

Let us first look at the phenomena that are due to the effects of the Earth's atmosphere and come into the realm of meteorology. Visible light from the Sun is made up of a full spectrum of colours. Earth's atmosphere scatters the shorter, bluer wavelengths more than red. Consequently blue light appears to come from all directions in the sky. Called Rayleigh Scattering, this is why the sky is blue.

When the Sun is low down at sunrise or sunset, its light must pass through a thicker layer of atmosphere. This both scatters away more blue light and dims the Sun's intensity. The light that's left gives us our red/orange sunrises and sunsets.

Green Rim, Red Rim and Green Flash

Another colourful effect sometimes seen with a low Sun is related to the dispersive properties of the atmosphere. Here, light passing through the ever-thickening layers of atmosphere as you get closer to the horizon spreads its component colours out like a giant prism. This can give the rising or setting Sun

this is best seen against a sea horizon. Visually, the green flash is difficult to see from Britain or the northern United States, although under magnification – for example using a camera with a telephoto lens – a distinct green edge and detachments can often be seen during clear horizon sunsets. Photographing the Sun in this manner should be done using a camera's review screen and not by direct viewing through the camera's viewfinder.

Halos

Sometimes, generally when close to full, the Moon appears to be surrounded by a luminous ring. This is called "a lunar halo", and is due to the moonlight being bent or refracted by very thin, high clouds known as cirrostratus. They are well over 20,000 feet above the ground, and are composed of ice crystals. They bend light at an angle of 22 degrees, causing a lunar halo spanning 44 degrees in the sky. One can also have a solar halo for exactly the same reason. The actual cloud is so thin that with the naked eye you cannot see it at all.

a coloured edge that appears green at the top and red at the bottom. The visibility of the rims is best picked up when the Sun is about 2 degrees above the horizon by using the solar projection method (see page 45) or photography. As usual, it's important not to look directly at the Sun.

Under certain atmospheric conditions, as the last edge of the Sun sets below the horizon, a green detached portion remains for a couple of seconds. Known as a green flash,

[1] Green rim (top) and red rim (bottom), visible at sunset.

[2] A 22 degree lunar halo.

[3] A primary rainbow with a weak secondary just visible to the right of the image.

[4] A lunar corona.

[5] Delicate colours visible in iridescent clouds.

Sun Dogs and Moon Dogs

A 22 degree solar halo can sometimes show brightenings to the east and west of the Sun. These are known as Sun dogs (parhelia) and on close examination show the colours of the rainbow. When seen either side of the Moon during a lunar halo, they are known as Moon dogs (paraselenae).

Rainbow

This is a commonly seen phenomenon caused by droplets of moisture in the Earth's atmosphere refracting (bending) different wavelengths of light by different amounts, smearing the wavelengths out into a spectrum of colours. Rainbows always appear in the section of sky opposite the Sun, and often indicate that rain is on the way or has just passed through. In a so-called primary rainbow, the effect is caused by light being refracted once in the droplets of water. In a double rainbow a second arc may be seen above the primary arc. This secondary rainbow is caused by light being refracted twice inside the water droplets. The sky in between the bows is darker and is called the Alexander dark band. At night-time, the Moon can also produce rainbows, though they are very much fainter than solar rainbows.

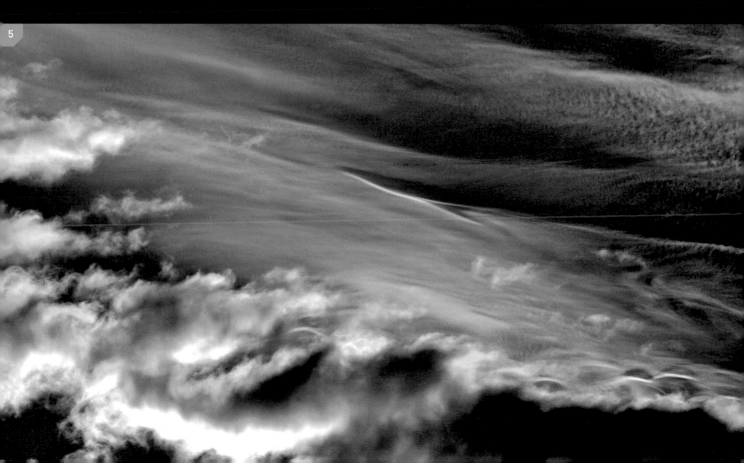

Water droplets or small ice crystals in a layer in front of the Sun or Moon can produce a palette of vivid pastel colours by an effect known as cloud iridescence.

Corona

A thin cloud layer in front of the Sun or Moon can give rise to a set of coloured rings known as a corona. More commonly seen around the Moon, this effect can be produced by any small particles, such as fine pollen grains suspended in the air. As pollen grains aren't necessarily spherical and are often aligned by the prevailing wind, a corona created by light passing through a pollen cloud may be elongated in shape and contain bright patches.

Crepuscular Rays

So-called "crepuscular rays" are rays of sunlight that appear to radiate from a single point in the sky. They are caused by sunlight streaming through gaps in the clouds or between other objects and are basically columns of sunlit air separated by darker regions of shadowed air. Even though crepuscular rays appear to radiate away from the Sun, they are in fact parallel to one another in the sky. Under rare circumstances, if you face in the opposite direction to the Sun you can see rays converging to a point. These are known as anti-crepuscular rays. Rays of light penetrating through holes in low clouds are often termed a Jacob's ladder.

Mirages

Under normal circumstances, the temperature of the atmosphere falls smoothly with height. This makes the Sun or Moon appear flattened when close to the horizon. Where warmer layers of air interrupt the otherwise smooth temperature fall, other effects can occur, such as the image of a distant object appearing above or below the real object. When the image appears below the real object, this is called an inferior mirage; when above, it's called a superior mirage. The effect of a shimmering reflective pool in the distance along a hot tarmac road is an example of an inferior mirage.

A *Fata Morgana* is an example of a superior mirage creating an image that sits above the real subject. These are commonly seen in polar regions above large sheets of flat ice.

A warm layer of air close to the surface of the sea can introduce some interesting effects on the rising or setting Sun. One particularly impressive display results when the Sun's image is reflected close to the horizon resulting in what looks like the Greek letter omega. An alternative name for this effect is the Etruscan vase.

[6] An Etruscan-vase sunset, also known as an "omega sun".

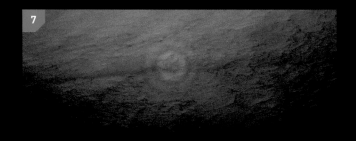

[7] A glory, photographed from the window of an aircraft.

[8] & [9] A display of night shining or noctilucent clouds over Selsey, England.

9

Glory

A glory is an optical phenomenon appearing in the clouds like a projection of the observer's head with a halo of light around it. It is produced by light scattered back from its source by water droplets. Glories can often be seen from mountains and tall buildings where there are clouds of fog below the level of the observer. They are associated with the "Spectre of the Brocken", the enormously magnified shadow cast by the low Sun on the under surfaces of clouds. The name comes from the Brocken, the tallest peak in the Harz mountain range in Germany where the phenomena is often seen. From an aircraft you can often see a glory with the aircraft's shadow in the centre.

Airglow

Airglow is a faint glow caused by various processes mainly in the upper atmosphere. During the day, scattered sunlight dominates but at night, from a clear, dark site, airglow can impose limits on the faintest objects that can be seen and photographed. The bluish glow from airglow may be seen visually in a band 10 degrees above the horizon where the atmospheric layer is thick. Lower down, extinction dims the glow, rendering it harder to see.

Noctilucent Clouds

All normal clouds belong to the lowest part of our atmosphere, known as the troposphere; the highest of these normal clouds are at altitudes of 20 thousand feet. There are rare clouds which occur much higher than this in a very thin layer located at altitudes between 47 and 53 miles, in the atmospheric level known as the mesosphere. At this height, these clouds are able to reflect sunlight when the Sun is below the horizon while "normal" tropospheric clouds remain in darkness. Sometimes seen in the summer months from latitudes 50–70 degrees north and south of the equator, these clouds can put on quite spectacular displays glowing with a beautiful electric blue light. As they shine at night they are known as noctilucent, or "night shining", clouds, often abbreviated to NLCs.

[10] The Moon illusion – seen against familiar objects, the Moon appears larger than when alone in the sky.

[11] The gegenschein visible in the dark skies over the VLT in Paranal, Chile (credit: ESO/Y. Beletsky).

[12] The Moon photographed in broad daylight.

[13] The majesty of the Aurora Borealis from Norway.

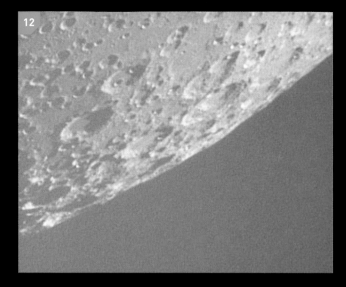

Moon Illusion

Most people over-emphasize the apparent size of the Moon in the sky, and in many paintings, artists often incorrectly show the Moon the size of a dinner plate. In reality the Moon's size is quite small, and its disc can easily be hidden behind your little finger held up at arm's length. However, when the Moon is low down and close to the horizon, it registers as being much larger than it does when it's high up, an effect first noted in ancient times. Ptolemy, the last of the great Greek astronomers (AD 120–180) had an explanation for it. He pointed out that when the Moon is low, it is seen across 'filled space' with trees and houses, while when it's high up, there's nothing to compare it with. This, according to Ptolemy, is why the low-down Moon looks much larger.

In modern times there are many explanations for the effect. One of the most plausible is that our brains interpret the sky as a flattened bowl above our heads rather than a sphere. When the Moon's on the horizon, our brains compensate for its perceived greater distance and we think it's larger than it actually is.

PM once tried an experiment on the beach at Selsey Bill, where he was joined by an old friend, Professor Richard Gregory. The full Moon was high in the sky and they rotated the image with a mirror while Richard noted the angles and measured the Moon's apparent size. They found that the Moon appeared much larger when low down, confirming the presence of the so-called "Moon illusion".

As an aside, the Moon's actual size is slightly smaller when it's closer to the horizon because it's an Earth radius further away and slightly squashed in height due to atmospheric refraction.

Gegenschein

The gegenschein is an elusive phenomenon that may be seen from sites with an extremely dark sky. The name in German means "counterglow" and refers to a faint patch of light that occurs opposite the Sun's position in the sky. This is due to interplanetary dust in the plane of the Solar System appearing fully illuminated in this direction. PM has seen it from Britain only once, in 1942 when the whole country was blacked out as a precaution against German air raids. However it is often seen in countries far from artificial light where the sky is completely clear.

[14] An auroral band cutting across Orion.

[15] & [16] More shots of the Aurora Borealis from northern Norway.

Moonbow

A moonbow is precisely the same phenomenon as a rainbow but is caused by the light of the Moon rather than that of the Sun as in a conventional rainbow. Of course it is much fainter, because the Moon is fainter than the Sun. How many full Moons would be needed to send as much light as the Sun? Many would guess five or six. The answer is close to half a million!

A moonbow is always seen in the part of the sky opposite the Moon, and the colours are so subdued that they cannot be seen with the naked eye. A moonbow is most easily viewed when the Moon is near full, and of course there must be rain falling opposite to the Moon.

Aurorae

One of the most beautiful sky phenomena is that of the aurorae or polar lights: Aurora Borealis in the northern hemisphere, Aurora Australis in the southern. Aurorae can be really magnificent, covering the entire sky and glowing with various colours. People who live in higher latitudes get the best auroral displays, while those living in temperate zones rarely see them at their best.

They are caused by electrified particles swept out from the Sun and crossing the 93-million-mile gap between the Sun and the Earth. After a rather complex interaction, charged particles eventually spiral down the magnetic field lines that channel towards the Earth's north and south magnetic poles. Here these

particles pass through our atmosphere and transfer energy to some of the atmospheric atoms they meet. This causes the atmospheric atoms to glow, resulting in the aurora. The atmosphere consists mainly of oxygen and nitrogen, and it's these elements that give the aurora its characteristic green, red and sometimes blue colour.

The aurorae occur in rings or rather ovals around the Earth's magnetic poles. It's not often that one has a spectacular auroral display from Britain or the United States (excluding Alaska!) but it can happen.

Remember of course that we are dealing with the magnetic poles and not the rotational ones, the two being somewhat separated. As the aurorae occur in ovals around the magnetic poles, the poles themselves are not the best places to see a display.

Northern Norway, for example, is right under the auroral oval and any dark night would seem very drab without its polar lights. Northern Scotland sees a good many aurorae per year, whereas in Sussex, where both authors live at present, they are very rare indeed.

Much depends on the Sun's activity and there is a fairly well-defined visual solar cycle of 11 years. Every 11 years the Sun is really active, with a disturbed surface showing dark patches known as sunspots, and this is a good time to look for aurorae.

Sunspots tend to occur in groups known as sunspot groups or active regions. These represent regions of intense magnetic energy. A highly energized active region may harbour enough energy for flares, the result of which is an enormous outpouring of energy. A powerful flare can produce a cloud of charged particles that can be sent out through the Solar System, known as a coronal mass ejection or CME. If one of these happens to head our way, it can disturb our magnetic field, creating what's known as a geomagnetic storm. This can also cause the auroral ovals to grow in size and spread further south, producing a bright display that may be seen at lower latitudes than normal. The Sun is dealt with in more detail in Chapter 4.

Although they are very beautiful, the magnetic disturbance behind a bright auroral display can disrupt communications and induce currents in long runs of overhead cables.

Zodiacal Light

Aurorae then, due to the Sun, are purely astronomical phenomena. So is a much more elusive phenomenon, the "zodiacal" light. Under very good clear conditions, after sunset, a cone of light may be seen extending up from the horizon, lasting for some time, and occasionally appearing quite bright. This zodiacal light is due to the Sun's influence on particles spread along the main plane of the planetary system. It does not last for long after sunset or before sunrise, because the sky has to be completely clear, with the Sun some way below the horizon. From the latitude of the Canary Islands, for example, zodiacal light is by no means rare, but it is very elusive indeed to see from Britain or the northern United States.

[17] The Zodiacal Lifght photographed from Tenerife, courtesy of Dr Brian May

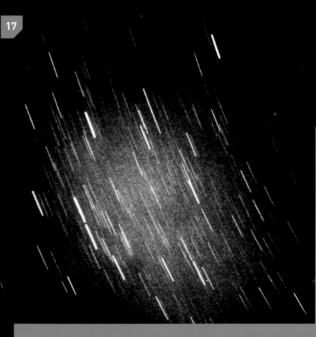

ASTRONOMY WITH YOUR CAMERA

Camera lenses, courtesy Bill Ebbesen.

Two of the most common questions from modern newcomers to astronomy are "Which telescope should I buy?" and "How do I take pictures of the night sky?". The second question is relatively new and is the direct result of the success of the digital camera, a device that has revolutionized astrophotography over recent years. For many, the lure of being able to capture beautiful images of the stars and planets is difficult to resist.

The function that allowed this to happen was the introduction of instant image review, a simple yet powerful facility that allows you to see what your camera has recorded as soon as an exposure is complete. With older film technology, the same process required the film to be removed from the camera and developed.

The instant review creates a rapid feedback loop allowing you to take a shot, review it and make the necessary adjustments to the camera settings to correct any issues identified during the review.

Apart from the obvious benefits of being able to correct poor setting choices or framing errors immediately, the long-term benefits of interactive digital imaging also help nurture a better feel for your camera's settings and what they actually do.

As an added bonus, most camera images store important setting values in the file header of the image. Should you forget what settings you used for a certain shot, it is therefore possible to use a simple image utility to read and remind yourself of these values.

[1] A digital single lens reflex (DSLR) camera.

Digital Sensor Basics

The term digital camera encompasses many different types of device but they all have one thing in common: an image sensor. The image sensor can be thought of as a grid of light-sensitive "photosites". When light falls on to a photosite, a charge is created and stored; the more light that a particular photosite encounters, the higher the charge.

At the end of an exposure, the grid of photosite charge values are read off and converted to numbers. Interpreted by a computer, these numbers represent the tonal image that the sensor saw, and reconstructed in the same grid formation, can be used to reconstitute the image.

The amount of charge a photosite can hold is called its "well depth". The interpretation of the amount of charge held is done by defining zero charge as 0 and a full well by a value equal to 2^n-1 (2^n means 2 multiplied by itself "n" times where n can be 1, 2, 3, 4, etc.), n being known as the sensor's bit-depth. The higher the value of n, the greater the maximum number becomes, and this translates into a greater number of tones in the final image. If n were trivially small, at say 2, a photosite's charge would be able to represent values between 0 and 3. The value of 0 is used to represent black while the maximum number represents white, so for n=2 the sensor could represent black, white and two tones in between.

A bit-depth of 2 would be very tonally limited, and modern sensors have far higher values of typically 8, 10, 12, 14 and 16. A bit-depth of 8 gives 256 tones including black (0) and white (255), while a depth of 14 gives 16,384 tones including black (0) and white (16,383).

As stated earlier, the amount of charge in a photosite's well is determined by how much light has fallen on the photosite. It is possible for a well to fill before an exposure has completed and this can lead to two conditions. In some sensors the well will fill and overflow, its charge cascading into adjacent wells. The architecture of an image sensor normally dictates the direction of flow to be constrained to the next well in the same column of photosites. The effect on the final image is that a line artefact appears from very bright sources (e.g. a bright star). Most general-purpose sensors avoid this issue by the application of an anti-blooming gate (ABG) to each photosite. This effectively caps a well when it gets full, preventing the occurrence of the cascade artefact.

[2] Colour images are produced from a greyscale sensor by overlaying a filter grid known as a Bayer matrix.

2

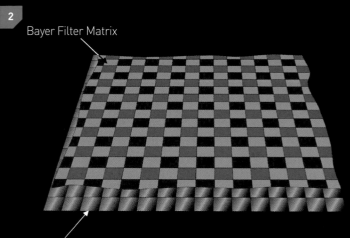

Bayer Filter Matrix

Sensor Photosites

One downside to ABGs is that when a star is over-exposed because the well receiving its light becomes full, it is not possible to make accurate measurements of the star's relative brightness beyond that point. As a consequence, sensors fitted with ABG technology are not useful for scientific brightness measurements unless the image remains under-exposed.

Colour

Imaging chips work as greyscale devices producing images which contain black, white and the grey tones in between. Although there are alternatives, most colour cameras commonly produce colour results by virtue of a Bayer colour filter matrix placed over the image sensor. This places a red, blue or green filter over each photosite and is arranged in a repeating pattern of 2 x 2 containing two green, one red and one blue filter. The resultant greyscale image is processed by a de-Bayering routine, which analyses the relative values that each photosite contains and compares this to its neighbours. With this information and knowing which colour filter was covering which photosite, the colour information of the scene can be restored.

Image Files

Once the photosite values have been read off a sensor and passed through a de-Bayering routine, the resulting data is stored in an image file. The photosite data that finally emerges from a colour camera has information on the tone and colour of that part of the image. These values are what are known as picture elements or pixels.

Image pixel data and other pertinent data are stored in an image file. The make-up of an image file is determined by the file format and there are many different types of these in existence. At the most basic level, an image file will contain an image header which holds information about the image, the camera settings used to take it and how the image data has been stored. The bulk of the image file will be the image data itself. In order to reconstruct the original image from the image file, a computer program that's aware of the file format will read the header and interpret the image data accordingly.

Image data can be stored in an uncompressed or compressed way. Uncompressed image-data files can be quite large so compressed formats tend to be popular, especially as digital imaging can quite easily generate thousands of images every year. There are two types of compression used – lossless and lossy. Lossless preserves all of the original image data while lossy uses special algorithms to remove selected information from the original image in order to reduce data. While this loss of information doesn't normally make a huge visual impact on the initial reconstruction of the image, such images are no longer viable for scientific analysis. Opening a lossy compressed

image and resaving it using a similar lossy format increases the loss of data from the original. Over several open-save cycles, the visual impact of this action can become noticeable.

Common Image File Formats

JPEG (extension .jpg, uses lossy compression)
Joint Expert Photographic Group format. Commonly used and adopted by general-purpose digital cameras. Level of compression can be defined. JPEG is limited to 8-bits-per-channel colour.

JPEG2000 (extension .jp2/.jpx, lossy and lossless compression)
An update to JPEG with greater storage flexibility. Not as common as JPEG due to greater computing requirements to decode/encode the format.

TIFF (extension .tif, typically lossless compression but can also support no compression and lossy compression).
The Tagged Image File Format isn't strictly a format in its own right but rather a transport framework for image data. The header of a TIFF file contains information on how the data is stored. Data may be uncompressed or compressed. The type of compression may vary too.

GIF (extension .gif, lossless format)
Graphics Interchange Format (GIF) images were once the mainstay of the internet. Supporting 8-bit colour, GIFs were also capable of providing animation capabilities which made them popular for website design. GIFs are still used in astronomy for simple "flick-book"-type image animations.

PNG (extension .png, lossless)
The Portable Network Graphic format is a popular compressed format used for internet images. It originally came into being to replace the commonly used GIF format which became subject to licensed use.

FITS (extension .fit, normally uncompressed)
The Flexible Image Transport System, or FITS, defines an information storage standard commonly used for astronomical images and data. Cooled astronomical CCD cameras often store their imaged results in FITS files. The FITS standard, like TIFF, is flexible and extensible. Many dedicated astronomical image processing programs can read FITS files and many general image editors can open FITS files via plugins. As well as image data, a FITS file can contain extra data pertinent to the image itself. For example, the temperature of the CCD chip can be recorded and this data used by certain programs to match the image with calibration files taken at the same temperature.

Camera Raw (extension depends on camera manufacturer, uncompressed)
Many high-end cameras will allow their images to be stored in a RAW uncompressed file which can be read by proprietary software or possibly by general image editing programs if the format is from a popular make of camera. These files are, in theory, untouched by the camera's internal tweaking routines and represent the purest form of image that the camera can deliver. RAW files are popular with deep-sky imagers for this reason.

AVI (extension .avi, compressed or uncompressed)
The Audio Video Interleave format is, as in the case of TIFF files, a transport framework for video data. As such it can employ a number of different ways for storing video data, which may be compressed or uncompressed. The mechanism used for storing the data is defined by the program that encodes and decodes the video data, known as a codec.

Image Size

The size of a reconstructed image file is often stated in terms of the number of pixels it contains horizontally and vertically. Multiplying these figures together gives the total number of pixels in the image and indicates the "resolution" of the

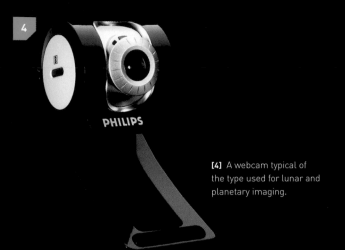

camera's sensor that took the image. This value is often stated in millions of pixels or mega-pixels, abbreviated to Mp or Mpx. For example, an image size of 3,504 x 2,336 pixels would contain 8,185,344 pixels or 8.2 Mpx. The number of image sensor photosites is also typically represented by a megapixel count.

CMOS vs CCD

Digital cameras typically use one of two types of imaging chip – CCD or CMOS. CCD or Charge Coupled Device sensors use photosites which convert incoming light into charge. CMOS or Complementary Metal Oxide Semiconductor sensors have additional circuitry next to each photosite which converts incoming light to a voltage.

Both types have pros and cons but the main objection to the newer CMOS technology when used for astrophotography came about because of higher noise levels. However, as the technology has matured, the gap between the two types has decreased.

Types of Digital Camera for Astrophotography

So far we have described the fundamental basics of digital sensors, the device that sits at the heart of a camera. Now we'll have a look at the cameras themselves. The number of digital cameras that are able to take astronomical photographs is quite large and in some cases rather surprising.

Photographic Digital Cameras

Photographic digital cameras are cameras that are normally used for general photographic purposes but which can also be coerced into use for astrophotography. There are a lot of devices which fall under this banner, from basic point-and-shoot models through to highly sophisticated digital single lens reflex cameras (DSLRs) which, as we'll see below, are capable of taking over the role that was so well occupied by film cameras not so many years ago.

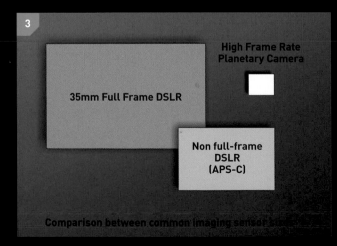

High Frame Rate Planetary Camera

35mm Full Frame DSLR

Non full-frame DSLR (APS-C)

Comparison between common imaging sensor

[3] Size comparison between common astronomical camera sensors.

[5] A dedicated planetary imaging camera.

[6] A cooled astronomical CCD camera.

Camera Phones

Functional convergence now means that many models of mobile phone have digital cameras built in and many of these are capable of taking basic astronomical photographs of bright subjects.

Webcams (High-Frame-Rate Cameras)

There are certain models of webcam which can be used for high-frame-rate imaging of bright Solar System objects. A bit of modification is normally required to fit them to a telescope, and for many years, certain models such as the Philips SPC-900NC produced images that were impressive to say the least. Such models are now becoming harder to locate, the role of the humble webcam now being usurped by more expensive and dedicated planetary imaging cameras. To control and use a webcam, it must be connected to a computer which, for portability, is normally a laptop.

Planetary Imaging Cameras (High-Frame-Rate Cameras)

Despite their name, as well as being able to take pictures of the brighter planets, planetary imaging cameras can also be used to image the Moon and, when a suitable filter is fitted, the Sun.

They can be thought of as a robust form of webcam, designed to capture short-exposure still images at impressively high rates. The current crop of cameras at the time of writing (2012) are able to record sequences of still frames at up to 120 frames per second. Like a webcam, a planetary imaging camera needs to be connected to a computer.

Cooled Astronomical CCD Cameras

Long exposures using digital sensors do have a tendency to suffer from noise – basically features in an image that shouldn't be there. There are many forms of noise and some are easier to deal with than others. Random thermal noise is one of the harder ones to get rid of and is caused by the generation of false signals due to thermal effects in the imaging chip. One way to suppress thermal noise is to cool the imaging chip well below the ambient temperature. There are various ways to do this, but the most common is to use a cooling circuit which exploits the Peltier effect, a thermoelectric cooling effect. Bringing the image sensor's temperature down by tens of degrees Celsius can reduce the effects of thermal noise to a minimum. Cooled astronomical CCD cameras use this technique. Unlike a stand-alone general-purpose camera, these cameras are designed to be connected to and controlled by a computer.

[7] Integrating video camera used for low-light astrovideo.

[8] A DSLR camera adapter ring.

[9] A point-and-shoot camera mounted on an afocal coupling platform.

Camera Summary

Camera Type	Typical Targets
Mobile phone camera	Low-resolution and wide-field Moon, bright planets and filtered Sun
Point-and-shoot camera	Landscape shots, constellations, general stars
High-end point-and-shoot (prosumer) camera	Landscape shots, deep sky, full-disc Moon
DSLR	Deep sky, landscapes, full-disc Moon, full-disc filtered Sun, general stars and constellations
Webcams	High-resolution Moon, bright planets and filtered Sun
Dedicated high frame rate planetary camera	High-resolution Moon, bright planets and filtered Sun
Cooled CCD camera	Deep sky
Astronomical video camera	Wide-angle night sky (e.g. meteor showers, satellites), Moon, filtered Sun, bright planets, deep sky

Astronomical Video Cameras

Low-light video cameras are now available for astrovideography, the process of taking video footage of astronomical objects. These can be fitted into the eyepiece barrel of a telescope producing a "live" image of what the telescope can see. These are especially good for group activities where a procession of observers waiting to look through the eyepiece of a telescope needs to be avoided.

A widefield video lens fitted to these cameras produces a view of large areas of the night sky which are far deeper than you can see with just your eyes. This makes video astronomy particularly good for meteor showers, where the whole shower can effectively be recorded and played back at your leisure.

Cameras and Lenses

In order for a digital camera to perform its primary function, focused light must be delivered to its sensor. This is achieved by fitting a lens to the front of the camera. Some cameras, such as those found in camera phones and point-and-shoot models, have fixed lenses fitted. Others, such as DSLRs and astronomical CCD cameras, are camera bodies on to which a variety of different lenses can be fitted.

Whether or not the lens can be removed from your camera determines the type of imaging you can do. Fixed-lens cameras have less natural flexibility, being constrained to take wide-angle sky shots the size of which is determined by the focal length of the attached lens. They can also be used in conjunction with a telescope for afocal imaging, the basic technique of pointing the camera down the eyepiece. This works for bright objects such as the Moon, filtered Sun and bright planets. Afocal imaging can also be performed successfully with certain camera phones.

Afocal imaging can be done by hand holding the camera in place but it's not easy to make sure the camera is focused, set correctly and pointing squarely down the eyepiece. In addition, pressing the shutter button can create unwanted movement, blurring the end result. For the best afocal images, it is necessary to hold the front surface of the camera lens as close as possible to the outward lens surface of the eyepiece but it's also important to avoid direct contact as glass on glass will almost certainly cause scratches.

One solution to this is to use an afocal coupling platform, available from many astronomical stockists. This locks on to your telescope's eyepiece holder and keeps the camera securely in place, providing the means necessary to adjust the camera's position for optimal results. A remote shutter-release cable can then be used to activate a camera's shutter without direct contact. As an alternative, a camera's shutter-delay timer can also be used to the same effect, giving a few seconds for any touch-induced wobbles to disappear before the shutter opens.

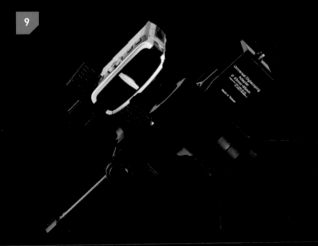

[10] DSLR camera prime-focus coupled to a telescope.

DSLR cameras offer much greater flexibility because, as well as being able to support a range of different lenses, they can also direct-couple to a telescope, effectively using it as a large telephoto lens. A DSLR is normally, but not exclusively, attached to a telescope using prime-focus coupling. This method uses the telescope as the main lens for the camera and no intervening camera lens or eyepiece is used.

All that's required for prime-focus coupling a DSLR camera to a telescope is an adapter ring specific to your camera make and model, together with an eyepiece barrel; both should be readily available from astronomical stockists. The ring locks into the camera body as would a normal lens and presents a female T-threaded aperture. The T-thread is a commonly used thread for joining various optical devices together in astrophotography.

T-thread:	42 mm x 0.75 mm pitch (sometimes referred to as M42 x 0.75)
M42 thread:	42mm x 1.00mm pitch

The eyepiece barrel should have a matching male T-thread which can be screwed into the camera adapter. The end result is a camera body that can be slipped into the telescope eyepiece holder and locked into position. For normal DSLR use and if your telescope supports it, a 2-inch barrel is to be preferred over a 1.25-inch one so that the imaging chip can be properly illuminated without light being clipped by the edge of the barrel.

Webcams normally come with a lens pre-fitted. For those models which can be used for astrophotography, this lens needs to be removed and replaced by a 1.25-inch eyepiece barrel. Astronomical stockists carry adapters designed to insert into certain webcam bodies. It's recommended to purchase an adapter that is pre-threaded to accept 1.25-inch filters. In this way, it's easy to add an IR-cut filter, sometimes known as an infra-red blocking filter, for better end results. A 1.25-inch adapter is normally fine for a webcam chip which will typically be much smaller than that used in a normal photographic camera. High frame rate planetary imaging cameras normally come with their own 1.25-inch adapters.

Cooled CCD cameras typically come with pre-threaded bodies presenting either female T-threads or the slightly different pitched M42-thread. Although it's possible to partially screw one of these threads into the other, care has to be taken not to go too far as the pitches used (that's the number of threads per unit distance) are different and damage can occur. Telescope eyepiece barrels, normally supplied with the camera, may then be screwed into these threads, allowing the camera to be inserted into the telescope's eyepiece holder. It's also possible to screw correctly threaded camera lenses into CCD cameras, giving access to large areas of the sky. The M42 thread, for example, was popularized by Pentax and there are many inexpensive Pentax M42 threaded lenses on popular online auction sites which can be used with M42-threaded CCD camera bodies.

[11] A beautiful photograph of the Elephant's Trunk Nebula (VdB 142) taken by amateur astronomer Ian Sharp.

CHOOSING AND USING A TELESCOPE

A long exposure of Pete Lawrence's telescopes taken in moonlight. Pete was carrying a torch while the exposure was being taken and this appears as a ribbon of light connecting the scopes.

CHAPTER 3

Having sampled what the sky has to offer to the unaided eye, or experimented with capturing images of the Moon, planets or stars with a digital camera, the next step is to purchase a telescope. This is not a decision to be taken lightly, and this chapter is designed to provide the essential information that you will need to choose a suitable instrument.

Astronomical telescopes are of two main kinds, refracting telescopes, or refractors, and reflecting telescopes, or reflectors. A refractor collects the light from its target object and passes it through a glass lens (also called an object glass or objective) where the rays of light are collected and brought into focus. The image is enlarged by a second lens known as the eyepiece or ocular. It is obvious that the larger the main object glass, the greater the amount of light that can be collected. This in turn means a higher maximum magnification can be achieved. Different eyepieces can give various magnifying powers. An astronomical refractor gives an upside-down image which, if required, can be turned the right way up with the addition of a correcting lens. Since this slightly reduces the amount of light reaching your eye, correcting lenses are rarely used for astronomical purposes.

One disadvantage of a basic refractor is that the different parts of the spectrum of light are brought to focus in different planes, so a bright image tends to be surrounded by gaudy coloured rings that may look beautiful but are a nuisance to the astronomer. This difficulty is avoided with a reflecting telescope, where the light collection is done by means of a mirror. A popular reflector is the Newtonian, first designed by the great scientist Isaac Newton way back in 1651. The light falls on a mirror which is specially shaped and reflects back on to a smaller flat mirror held at an angle of 45 degrees. This mirror diverts the light to the side of the tube, where it is enlarged by an eyepiece as before. In a Newtonian reflector you look into the side of the tube instead of up it.

[1] Patrick Moore's first telescope. The classic 3-inch brass refractor which he has owned since boyhood.

[2] A modern computer-controlled 80mm (3.15-inch) starter scope.

[3] A long-focal-length telescope is best for getting close-up views of Solar System objects such as the crater Clavius shown here.

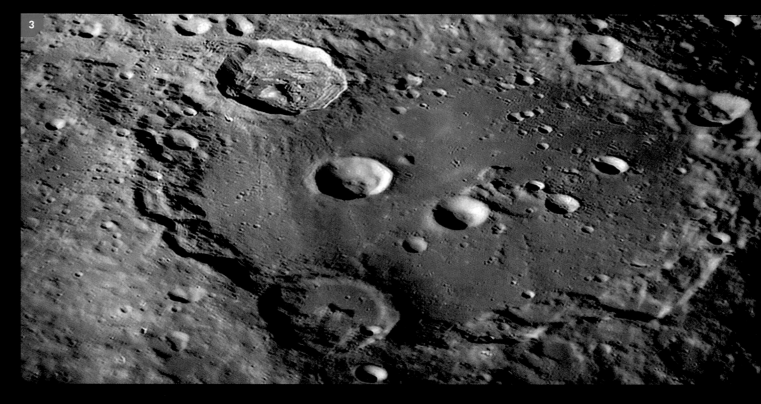

From the amateur's point of view, these two types have their own advantages and drawbacks. The refractor needs very little maintenance and will last a lifetime if reasonably well treated. On the other hand, a reflector has to have its mirrors coated periodically with some reflecting substance, generally a thin layer of aluminium. This has to be done regularly. In addition, a reflector is prone to going out of adjustment. Against this, a reflector is cheaper than a refractor of equal power because a large mirror is easier to make than a large lens; all the world's largest telescope are reflectors. The largest refractor, the Yerkes, is 40 inches across; the largest single mirror reflectors have mirrors more than 323 inches in diameter.

Modern manufacturing techniques have opened the market for new and improved, cost-effective telescope designs. These include large-diameter reflectors, colour-corrected refractors and hybrid telescopes known as catadioptrics, which incorporate both lenses and mirrors.

Choosing a suitable scope out of what has become a rather bewildering array of instruments has actually become more complicated over time because of the extended choice. In reality there is probably no one simple answer to the question of which telescope is best for you, but a simple bit of self-questioning may help to narrow your options quite considerably.

The first question is one of budget and how much are you prepared to spend? For a first scope purchase this is a pretty serious question. Do you spend lots of money on a top-of-the-range model which you'll keep and use for many years, or do you spend a small amount on a telescope to let you dip your toe in the water? The only person that can answer this question honestly is you, but experience suggests that the second choice is probably the right one. Spending a lot of money on what is essentially a blind purchase may lead to disappointment, an unused scope and wasted money.

If you are looking to upgrade from an existing scope to a more advanced one, the choice is easier because you will know about the foibles and limitations of your current instrument and have a greater sense of the direction you want to move in.

Once your budget has been set, the next decision needs to be geared toward what you want to look at in the sky. The choice ranges from the Moon and brighter planets to the stars, galaxies, nebulae and clusters that pepper the night sky.

Let's deal with the simple situation first, when your interest is either Solar System or deep-sky objects. For Solar System observing you will need a large scope with a long focal length. The larger the diameter of the telescope the more resolving power it will have, enabling you to see smaller and smaller detail. Unfortunately, with increased aperture size comes increased cost, so this is where your budget limit kicks in!

The brighter planets and the Moon require a reasonably high magnification to see significant detail. As magnification is calculated by dividing the focal length of the telescope by the focal length of the current eyepiece, it's clear that a telescope with a naturally long focal length will be able to get up to high magnifications more easily than one with a relatively short focal length. Similarly, if your interests are in imaging these objects, a long-focal-length instrument will give you a better opportunity to use higher magnifications.

As an example, consider a telescope with a 400mm focal length using a 10mm eyepiece. Here the magnification will be 400/10 = 40x. Compare this with a telescope with a 2000mm focal length using the same eyepiece; the equivalent magnification is 2000/10 = 200x, which is much better suited for planets.

You cannot go on increasing the power indefinitely though, and the rule of thumb for the maximum useful magnification you can achieve is obtained by multiplying the aperture in inches by 50. So for an 8-inch scope, the maximum useful magnification would be 400x, but this would only be useful if the atmosphere you were looking through was very stable.

The ideal telescope for looking at the Moon and planets would be a long-focal-length, large-diameter, colour-corrected refractor, but here the budget limit is easily exceeded with alarming speed. An alternative would be a large-diameter, long-focal-length reflector, but these can be quite bulky and difficult to mount and handle. Scopes that use folded optics, known as catadioptric telescopes, can offer a good compromise here, combining a relatively large aperture with a long focal length, all for more realistic prices.

At the other end of the range, a telescope designed to look at the stars and deep-sky objects needs be large and have a relatively short focal length. A large reflector fits this bill well, and there are plenty of examples of this "light-bucket" type of telescope available on the market. One very popular and relatively cheap version of a deep-sky light bucket is the Dobsonian reflector. This is typically a large-aperture telescope on a simple left-right/up-down (alt-az) mounting platform. The ethos behind this telescope, originally designed by John Dobson, was that this should be a telescope that was simple to make from readily available components. If you are good with your hands, there are many Dobsonian plans available online which would allow you to create a large-diameter scope for very little money.

The distinction between a planetary scope and the deep-sky scope is often made by referring to the speed of the instrument. This is a measure of the instrument's focal length divided by its diameter using the same units. So a 200mm (8-inch) reflector with a 2000mm focal length has what's called a focal ratio of 2000/200 = 10. This is normally written as f/10 and is an indication of the speed of the telescope, in this case f/10 being considered slow.

A telescope with a similar aperture of 200mm but a shorter focal length of 600mm would have a focal ratio of 600/200 = f/3, which is considered fast. If you're wondering where the terms "speed", "slow" and "fast" come from, they echo the days

[5] A big-aperture and low-focal-ratio (fast) scope is ideal for deep-sky viewing.

of film photography. The term "speed" is a measure of how quickly a lens or telescope can deliver a set amount of light to photographic film. A "fast" lens would achieve delivery quicker than a "slow" lens. Basically, a slow lens requires a longer exposure to achieve the same depth of image that a fast lens can deliver in a relatively short exposure.

The basic rule of thumb is that slow instruments are good for the Moon and planets while fast ones are better suited to deep-sky objects and the stars. What are the definitions of slow and fast? Well, this is open to interpretation but a reasonable assumption would be that any scope with a focal ratio of f/5 or lower is considered to be a fast instrument, while any scope with a focal ratio of f/9 or higher is considered to be slow. Of course this leaves those with focal ratios in the region of f/5–f/9 uncategorized but these are the instruments that are suited to both camps. If you regard your interests to be equally divided between Solar System and deep-sky objects, then the f/5–f/9 instruments are probably the ones that will interest you most.

One further factor when considering the speed and focal length of a potential purchase is the use of an optical amplifier

[4] If you're just starting out, a pair of 7x50 or 10x50 binoculars are ideal.

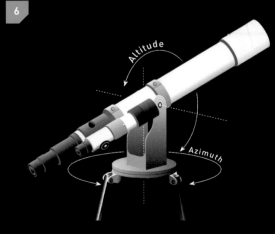

[6] A basic alt-az mount provides side-to-side and up-and-down motion.

to alter these figures. An amplifier, such as a Barlow lens, can be used between the telescope and eyepiece to effectively make it seem that the focal length of the telescope has been multiplied by the power of the Barlow. For example, consider a 200mm aperture telescope with a 2000mm focal length (f/10). Using a 2x Barlow increases the effective focal length of the scope to 4000mm but reduces its speed to f/20. The caveat here is that amplifiers with powers greater than 1 can make viewing harder, the high focal ratios produced giving rise to dimmer images.

Optical amplifiers also come with powers less than unity, called focal reducers. These reduce the effective focal length and increase the speed of the instrument, delivering a wider, brighter view.

Careful and considered use of optical amplifiers will mean that you can partially adapt a long-focal-length scope for use as a deep-sky instrument and vice versa. However, the results will not be as good as a dedicated scope.

To assist you further, below is a general indication of the best type of scope to buy for each budget group. It's by no means meant to be a definitive list but hopefully if you're in the market for a telescope, it will point you in the right direction.

Up to £100 This is probably too low a budget for a new telescope so the recommendation here would be to purchase a good pair of, say, 10 x 50 binoculars.

£100–200 There are numerous small refractors available in this price range with apertures 60–90mm. Avoid computerized control and concentrate on optical quality and a good mount.

£200–500 A 10-inch Dobsonian or equatorially mounted 8-inch reflector make great deep-sky telescopes. A 4-inch refractor is another option for lunar and planetary observing.

£500–1000 At this level look for good-quality optics on a robust equatorially driven mounting.

£1000 plus Optics and sturdy mount again come first. Colour-corrected refractors or catadioptric instruments for high-magnification views of the Moon and planets are recommended.

[7] Optical amplifiers such as Barlow lenses can be used to alter the effective focal length of a telescope.

Try Before You Buy

If you want to try out a particular telescope before you buy it, have a word with your local astronomical society to see whether anyone in the group has the same or a similar model that they are willing to let you look through and examine. If you have no luck here, have a word with your telescope stockist to see whether they know of anyone locally who may be able to offer the same service.

An internet search on a particular type of telescope, stating the manufacturer, type and size, is a good way to see whether there has been any user feedback about the instrument online. Here you can often pick up existing problems or even learn that the model you have chosen is regarded as a superb bargain.

Second-Hand Options

There are a number of options available if you are happy to purchase a second-hand telescope. There are numerous online second-hand sites where you may be able to pick up a bargain. eBay is another potential source. As usual when buying

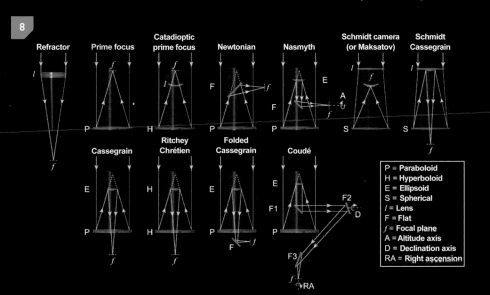

[8] Different telescope designs shown schematically. Yellow lines represent parallel light rays entering the scopes and being brought ino focus at f, the focal plane, which is where the eyepiece is located. l is the lens in the first image, of a refracting telescope. The rest are reflecting telescopes, and the labels P, H, E, S indicate the shape of the mirror bringing the light into focus, explained in the diagram. F stands for flat – a flat mirror.

P = Paraboloid
H = Hyperboloid
E = Ellipsoid
S = Spherical
l = Lens
F = Flat
f = Focal plane
A = Altitude axis
D = Declination axis
RA = Right ascension

[9] A small Newtonian telescope on a basic equatorial mount.

[10] Single-arm, computer-controlled mount with Go-To computer control.

[11] A long-exposure wide-field image, such as that taken with a camera on a fixed tripod, will show the stars trailing. The trailing is caused by the rotation of the Earth on its axis, the stars appearing to rotate around the celestial poles.

anything privately, be cautious. Ask the vendor whether they are willing for you to see or collect the scope if you do purchase it. Even if you don't intend to inspect beforehand, this will help to verify that there is actually a telescope at the other end!

Mounts

So far we've made little mention of telescope mounts despite these being a vital component to ensure that you get the best out of your purchase. Telescope mounts are analogous to speakers in a hi-fi system. After spending your money on a new amplifier, there can be a reluctance to spend a fortune on high-quality speakers. Buying cheap here may result in a sound quality which doesn't reflect the quality of the amplifier at all well.

Similarly, if you put a decent telescope on a wobbly, cheap mount, you'll degrade the quality of the view you get and your experience may be disappointing and rather frustrating. Some telescopes are sold with mounts included while others require you to buy the mount separately. Some scopes are also sold integrally connected to the mount.

There are many different makes of mount available but, fortunately, the number of different types is actually quite small. The simplest mount is called an alt-az mount. The term "alt-az" is short for altitude-azimuth and refers to the fact that the mount moves from side to side (azimuth) and up and down in altitude. A standard photographic tripod is an example of an alt-az mount.

The problem with a basic manual alt-az mount is that it does not naturally follow the stars across the heavens. If you think about it, placing a small telescope on a photographic tripod allows you to move the telescope horizontally and vertically. The

stars do not move horizontally parallel to the horizon unless you happen to be observing from the North or South Pole.

If you watch the stars on a clear night, they move in circular arcs centred on a stationary point in the sky known as a celestial pole. In the northern hemisphere the stationary point is known as the North Celestial Pole (NCP) and is located in the sky very close to the star Polaris. In the course of a day, Polaris describes a very small circle around the NCP.

If you take an alt-az mount and tilt it so that the azimuth axis points towards the NPC, a telescope fitted to the mount will, moving in what is now tilted azimuth, follow the curved arcs the stars naturally follow across the sky. This is the principle of the equatorial mount. If you add a motor drive on the "tilted azimuth" axis which turns the axis at the same rate as the Earth rotates but in the opposite direction, the mount is able to compensate for the rotation of the Earth and anything in the eyepiece of an attached scope will stay in view and not drift out of sight.

So to recap, so far we've described an undriven alt-az mount, a tilted version of the alt-az mount known as an equatorial mount, and a driven version of the equatorial mount known as a driven equatorial mount.

Putting a drive on the azimuth axis of an an alt-az mount isn't that useful, as the mount does not emulate the natural motion of the stars unless, as stated, you happen to be observing from one of the Earth's poles. However, putting a drive on both the azimuth and altitude axes allows an alt-az mounted scope to follow the stars if a computer is involved to emulate their motion.

Modern computerized mounts can offer some impressive functionality but it's wise to treat them with caution for lower-priced telescope packages. If you're paying a set amount for a computerized telescope package, it's useful to keep in mind that the computerization comes at a cost and may detract from the size and possibly the quality of the telescope that is supplied with it.

Go-To capability is offered on many computerized mounts. Here, as long as the mount knows where it is in the world, what time it is, and the direction in which it's currently pointing, you can select the target you're after from a huge database of objects. The computer then does the rest, slewing the telescope round so that it's pointing in the right direction. The term Go-To describes the action of selecting your target and pressing a button to "go to" it.

Although this may sound the perfect way for a beginner to find objects in the night sky, there are issues. For example, letting the computer do the work will not help you learn your way around the sky. Here, there's no substitute for a bit of practical map-reading using charts and the naked eye. If your goal is to learn where things are in the night sky, the use of a Go-To mount is best avoided or at least limited. Then there's the issue of the onboard picking database, which for a typical Go-To system is measured in the tens of thousands of objects. Many generic Go-To systems are designed for a large variety of telescopes so using the database with a small-aperture instrument will invariably result in having access to many objects that simply cannot be seen through the eyepiece.

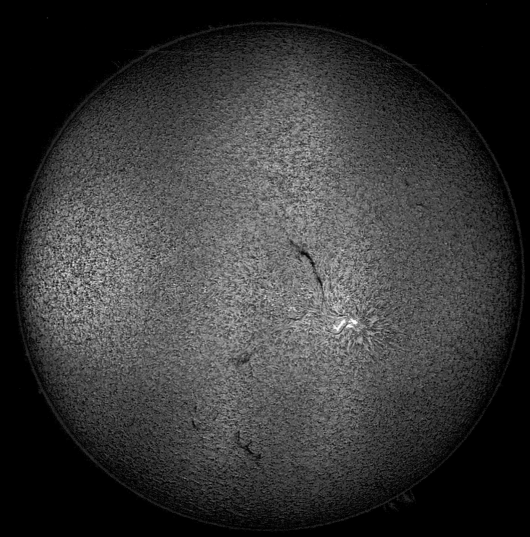

This is not to say that Go-To is all bad. If you know you're a casual observer who isn't that interested in learning your way around the constellations but would rather just "see" the items of interest that are up on any particular night, then Go-To is a good choice. Similarly, if you have very light-polluted skies so that navigating around the stars with the naked eye is tricky at best, then Go-To may help you out here too.

Go-To also works well for imagers where locating objects by eye can be wasteful of precious imaging time. And Go-To may be the only way to quickly locate objects that are below your scope's visual threshold.

Go-To systems can be used with alt-az and equatorially mounted telescopes. Computerized alt-az mounts are not ideal for imaging as they introduce an unwanted effect known as field rotation during long exposures. This makes the imaging target rotate around the central axis of the image frame.

Solar Telescopes

Generally, any one type of telescope can be used for viewing any type of object. Even if, as described above, the scope's focal length isn't ideal for the job, you can usually get some sort of view of the object you're after.

By fitting a white-light filter to the front of the telescope so that the entire aperture is protected, and capping or removing the finder, it's possible to turn a night-time telescope into an instrument capable of viewing the Sun in white light. However, this won't reveal much detail beyond what can be seen on the Sun's visible surface, the "photosphere".

For a look at the exotic features that inhabit the hot magnetically-influenced environment just above the photosphere, you need to use a special filter known as a hydrogen-alpha (H-alpha) filter. These are available as a filter set used to convert existing night-time telescopes for H-alpha viewing or as dedicated telescope packages. H-alpha filters do tend to be quite expensive, and a dedicated H-alpha telescope cannot be converted for night-time use; this can only be done if using a filter set. The cheapest dedicated H-alpha scopes currently (2012) retail for around £500.

Unlike their night-time equivalents, dedicated H-alpha scope apertures tend to be more restrained, the normal amateur range extending from 35mm up to 100mm. Larger amateur H-alpha scopes are available but price tends to keep their numbers rather limited.

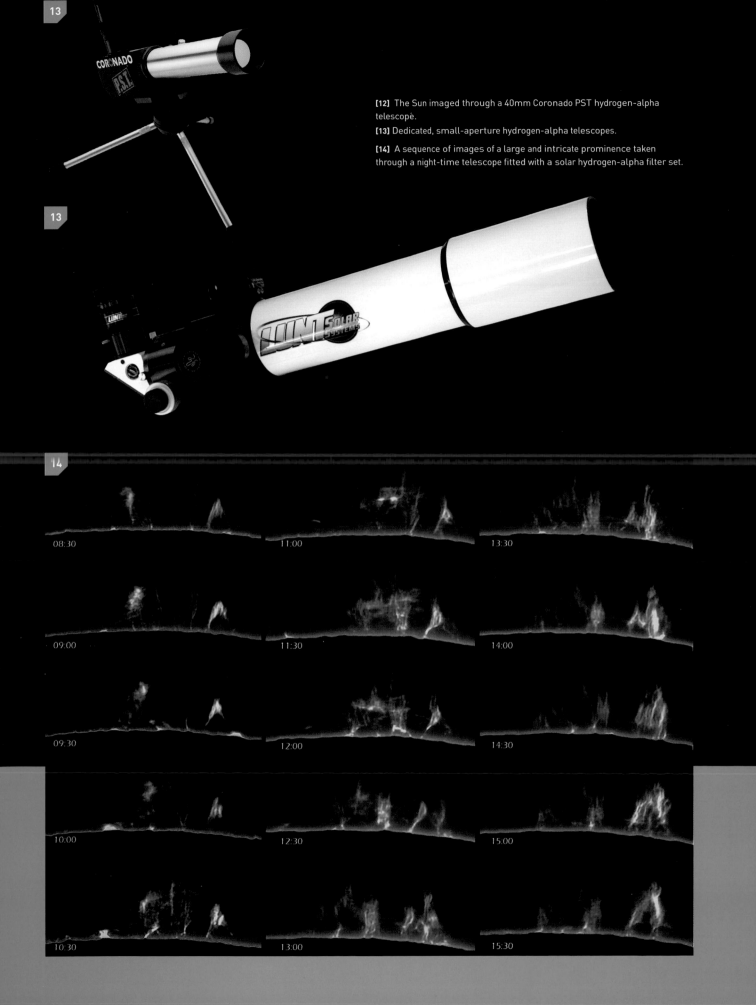

[12] The Sun imaged through a 40mm Coronado PST hydrogen-alpha telescope.

[13] Dedicated, small-aperture hydrogen-alpha telescopes.

[14] A sequence of images of a large and intricate prominence taken through a night-time telescope fitted with a solar hydrogen-alpha filter set.

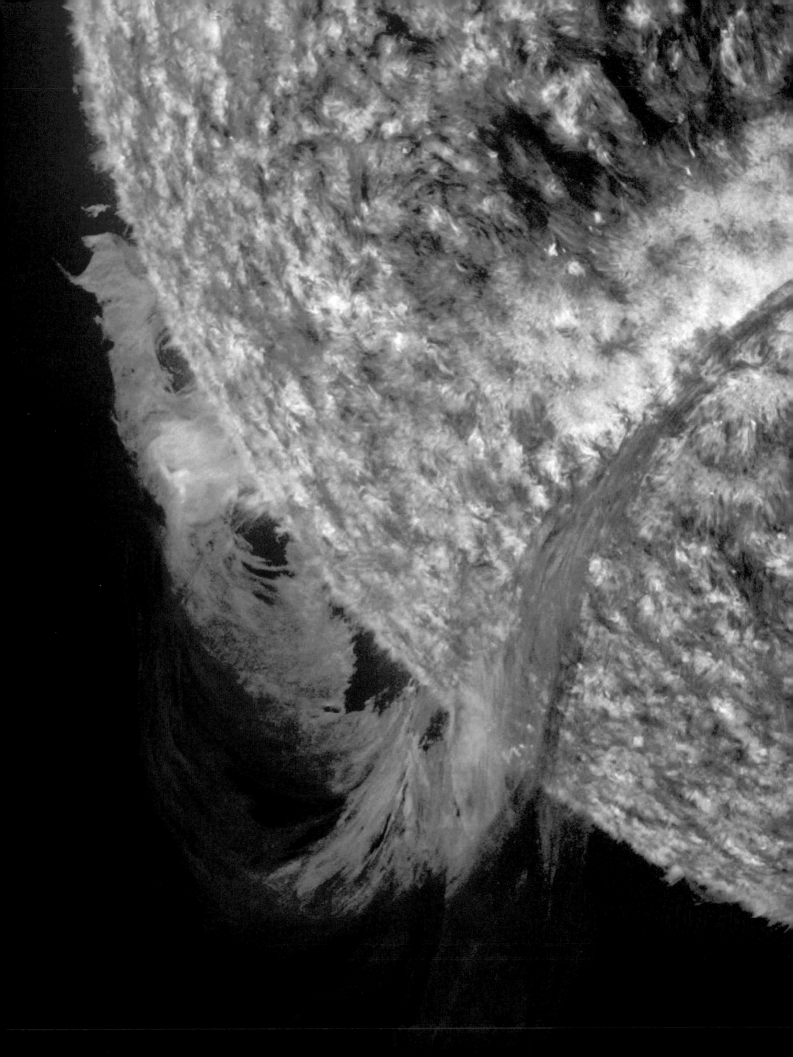

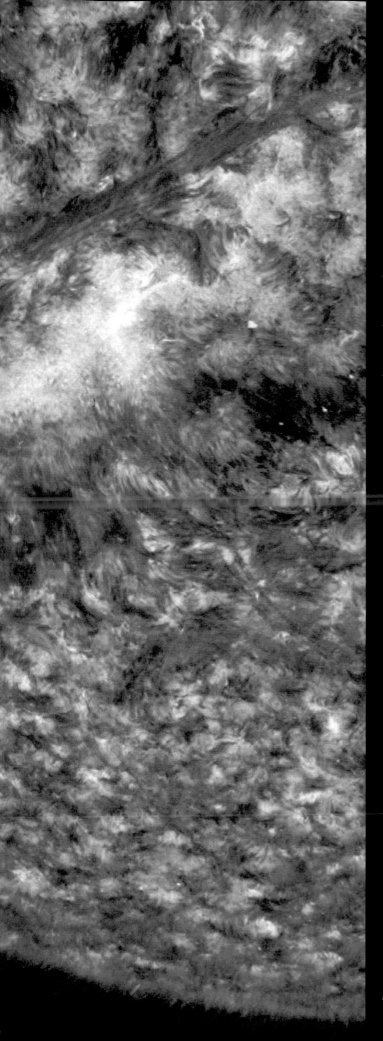

OUR STAR
THE SUN

A giant prominence appears off the edge of the H-alpha Sun, NASA/SDO.

CHAPTER 4

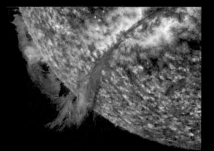

Many years ago, an astronomer named Proctor wrote a book called *Our Sun, Fire, Light and Life of the Solar System*. This is by no means a bad description. The Sun formed around 4600 million years ago, inside a gaseous nebula. As the nebula gases condensed, the Sun became hot and incandescent with a family of planets in orbit around it, of which the Earth is one. Solar observing is a fascinating pastime and does not involve the need for really large and expensive pieces of equipment.

We make no apology for stating the following warning as strongly as we can – never look directly at the Sun with your eyes or through any optical instrument unless your're using a certified, protective solar filter. PM remembers meeting a man in his 80s blind in one eye, who had blinded himself looking at the Sun using a dark filter over the eyepiece and was unable to get his eye out of the way when the filter cracked.

In some people's view, the safest way to observe the Sun is through a small refracting telescope, pointing the telescope at the Sun and projecting the Sun's image on to a screen held or fixed behind the eyepiece. This method is safe provided you keep your eye well away from the cone of light projected through the eyepiece. There is a real danger when using this projection method, however: never ever leave the telescope unattended. The natural inclination on seeing a telescope set up, especially for a child, is to look through it. If that telescope is pointing at the Sun, the result will be immediate and permanent blindness. Another method is to use a certified solar filter fitted over the entire open end of your telescope. This method is described in more detail later.

Sunspots

Viewing the Sun through an appropriate white light filter, you will sometimes see dark patches known as sunspots on its visible surface (photosphere). These represent regions of intense magnetic activity and look dark simply because they're around 1500 degrees cooler than the 6000 Celsius surrounding photosphere. Sunspots come and go according to an approximately 11 year visual activity cycle. At solar maximum there may be many spots, at solar minimum the Sun may be free from spots for days or weeks at a time. People often ask if sunspots have any effect on climate and weather. Well, all our light and heat comes from the Sun and slight changes can have profound effects. For example, between the years 1645 and 1714 there were virtually no sunspots and the solar cycle appeared to be suspended. This was the time of the so-called "Maunder Minimum" and the temperature was lower in Britain than now. The Thames froze every winter and frost fairs were held on it. Whether this was due to the Sun's activity or not is still the subject of debate, however in 1715 the spots returned and the situation reverted to normal.

We do go through periods of solar warming and cooling, and these are easily detectable though by no means catastrophic. Our own activities do have some effect, but how much is uncertain, and recently the topic has become increasingly political.

[1] & [2] Sunspots imaged through a 5-inch telescope protected by Baader Astrosolar film.
[3] Full-aperture solar filter.
[4] Solar image viewed by the projection method.

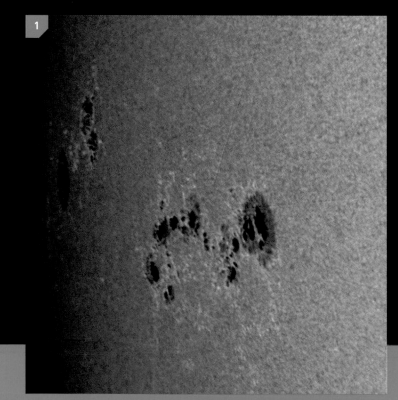

1

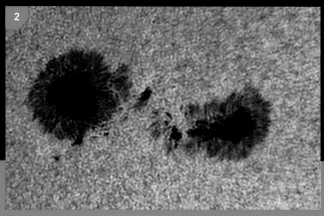

2

The Sun spins in a period of a few weeks, therefore any sunspots appear to travel across the disc from one side to the other and it is fascinating to track their change. A large spot consists of a dark central region or umbra surrounded by a lighter area or penumbra.

When a spot reaches the limb (edge) it is carried round to the far side of the Sun to reappear after a fortnight, if it still exists. A really large group may persist for several rotations, though small ones disappear after only a few hours.

You can learn a great deal about the Sun and its spots even with a small telescope, but for advanced studies we need more than a telescope. We need instruments based on the spectroscope, a device astronomers use to reveal the chemical make-up of stars.

Observing the Sun is a potentially dangerous activity and must be carried out with great caution and care. Projection methods should only be carried out with refracting telescopes. The use of specialist certified solar film allows you to use any type

of telescope as long as the full aperture of the instrument is securely covered. Note that cheap eyepiece Sun filters should not be used.

Astrosolar Filters

One of the most effective ways to protect a scope for white-light viewing or imaging is to construct a full-aperture filter for it out of a material called Astrosolar film made by Baader Planetarium. The film is available in A4-sized sheets or, for larger instruments, a 1.0 x 0.5m roll. Two versions of the film are available, in different filter densities, one for visual and one for imaging applications; the higher density (ND5) film is suitable for both uses. Astrosolar film is safe to use on any type of telescope. Simple instructions for creating a custom filter cell are given on the Baader Planetarium website. Other types of certified white-light filter are also available. The thin film nature of Astrosolar film means that it doesn't need to be, and indeed shouldn't be, stretched across the filter cell tightly. Ripples in the film have no significant effect on the incoming light path.

Before fitting, it's good practice to hold the filter up to the Sun for visual inspection. Any holes or tears indicate that the filter should be discarded and a new one made. If the filter passes visual inspection, it should be fitted to the front of your scope in a way that won't allow it to lift off easily. For safety, a piece of low-tack tape can be used to secure the filter to the scope.

Another important safety point concerns the telescope's finder, which is a small telescope in its own right. This needs to be capped, filtered or indeed removed before pointing the main scope at the Sun.

With the main filter fitted, and the finder dealt with, the next job is to point the telescope at the Sun. The safest way to do this is to look at the scope's tube shadow on the ground and position the telescope so that the tube's shadow profile is as small as possible. Once this has been achieved, it means the telescope will be pointing at, or at the very worst, close to the Sun. Final positional tweaks can then be done through the eyepiece.

The white-light Sun can be photographed in much the same way as the Moon and bright planets, using the set-ups described in Chapter 3 and elsewhere throughout this book. The best solar imaging results come from using a high-frame-rate camera because this provides a means to reduce the blurring and distorting effects of the Earth's atmosphere.

As the Sun heats the ground, it's fairly common to find the view becomes less stable throughout the day. This is evidenced by the edge or limb of the Sun appearing to "boil". The resolution of fine detail on the Sun's visible surface (photosphere) is heavily dependent on the overall stability of the view. This stability is described under the general term "seeing". The Mount Wilson Solar Seeing Scale, shown reproduced overleaf, lists the different types of solar seeing that may be encountered. An arcsecond, by the way, is 1/3600th of a degree or roughly 1/1800th the apparent diameter of the Sun in the sky.

5 = Solar image looks like an "engraving". Extremely sharp and steady. Limb motion and resolution 1 arcsec or better.

A high-frame-rate capture typically produces several hundred or even thousand still frames. In order to construct the final image, the best shots are extracted, aligned and averaged. There are various computer programs which can do this automatically such as the freeware AVIStack and Registax.

The obvious targets for white-light imaging are sunspots and sunspot groups. These high-contrast features are also useful as focusing targets. When imaging the Sun, start out using a low power first using prime focus. If the conditions are stable, optical amplifiers such as Barlow lenses can be used to boost the magnification but it is important to match the power to the conditions. Use too high a power under poor seeing and what you'll end up with will be fuzzy and low on detail.

If there are few or no sunspots visible, the edge of the Sun can be used for focusing. The outer regions of the Sun's disc in white light look darker than the middle due to an effect known as limb darkening. Look carefully in the shaded edge regions because here you may see lighter patches known as faculae. These are regions where strong magnetic fields reduce the density of gas in the photosphere, allowing us to see deeper into the Sun where hotter gas radiates with greater intensity.

The photosphere itself is textured. This can be seen visually with a 6-inch properly filtered scope or imaged with a filtered 4-inch under good seeing conditions. This texture, often referred to as a "rice grain" pattern, is known as solar granulation and

1 = Solar image looks like a "circular saw blade". Completely out of focus. Limb motion and resolution greater than 10 arcsec. Smaller sunspots will not be seen.

2 = Solar image is always fuzzy and out of focus. No sharp periods. Limb motion and resolution in the 5 to 10 arcsec range.

3 = Solar image about half the time sharp and half the time fuzzy. Some short periods where granulation is visible. Limb motion and resolution in the 3 arcsec range.

6

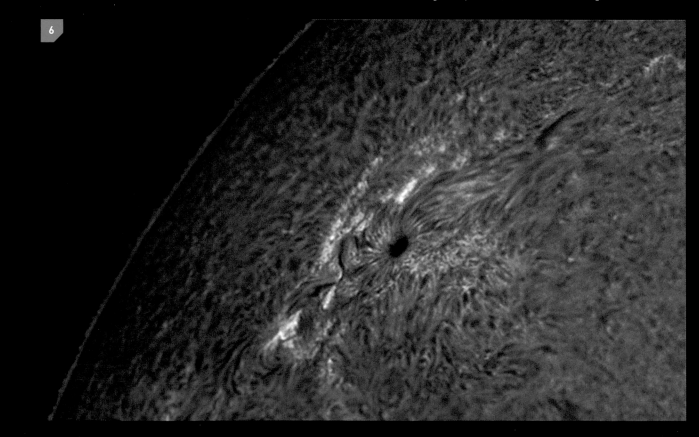

[5] Sunspot groups visible on the white-light Sun.

[6] A sunspot imaged through a hydrogen-alpha filter.

[7] Bright regions called plage and dark, snake-like filaments abound in this hydrogen-alpha image.

[8] & [9] Active regions of the Sun seen in hydrogen-alpha.

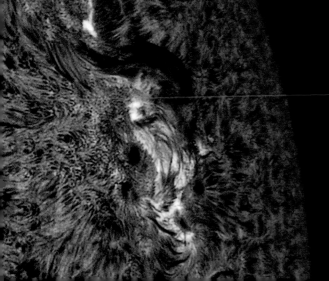

[10] A dedicated Calcium-K telescope.

[11] Hydrogen-alpha filter etalon.

[12] Filaments rising from the surface of the Sun.

[13] The Sun imaged through a Calcium-K telescope.

is due to the photosphere being made up of the tops of vast convective cells that start many thousands of miles below. Each "cell top" is approximately 600 miles across and typically lasts for 10–20 minutes before dissipating.

Narrowband Filters

The use of narrowband filters, such as those tuned to the hydrogen-alpha (H-alpha) wavelength of 6562.8Å (1Å = 0.0000000001m) or Calcium-K (CaK) wavelength 3933.7Å, reveal a completely different view of the Sun. These filters ignore the vast majority of the incoming light, concentrating on a very narrow window of wavelengths around the central value. Although these aren't the only narrowband filters available for solar work, H-alpha and CaK are by far the most popular. H-alpha can be used for both visual and imaging applications while CaK filters tend to produce an image which is difficult to see visually. As such CaK filters tend to be used for imaging only.

WARNING

It's important not to confuse deep-sky and solar H-alpha filters. Although they have the same name, deep-sky H-alpha filters are unsuitable and indeed dangerous for solar work.

Hydrogen-Alpha Filters

In order to work, a solar H-alpha filter needs to be manufactured to incredibly tight tolerances. Centred on the H-alpha spectral line, the window of extra wavelengths that the filter passes – known as the filter's bandpass – determines just what the filter will reveal. Typically, a solar H-alpha filter's bandpass will be in the order of 1Å or less. The central optical component of most solar H-alpha filters is an optical resonance cavity known as a Fabry-Pérot etalon. The incredible tolerances needed to make this component work are responsible for the often rather high price tag of these instruments.

H-alpha filters are available as sets for converting astronomical telescopes for H-alpha viewing or as dedicated solar telescopes.

A hydrogen-alpha filter reveals clouds of glowing hydrogen, which exist immediately above and beyond the photosphere. The photosphere itself is hidden in this view, beneath a blanket layer of hydrogen known as the chromosphere, approximately 6000 miles deep. The fine "rice grain" texture of the photosphere is replaced by a coarser pattern that resembles the skin of an orange, due to the presence of "dark mottles".

H-alpha features are heavily influenced by the Sun's magnetic field and this is particularly evident around active sunspot regions. Ironically, the high-contrast view of a sunspot under white light is lost in H-alpha, with significant detail being seen in and around the group itself. Of particular note are bright patches associated with active regions, known as plage.

H-alpha filters reveal some of the most dynamic features of our nearest star including vast clouds of hydrogen held above the chromosphere. When seen on-disc, these appear as dark snaking filaments. As the Sun rotates, these features may eventually reach the Sun's limb where they appear to hang off the curved edge, bright against the black space beyond. In this state, they are known as prominences.

H-alpha filters are also very good at revealing the violent outbursts from the Sun's surface known as flares. A powerful flare may hurl material off the Sun completely, sending it through the Solar System as a coronal mass ejection or CME.

CaK Filters

CaK filters produce an image in a part of the spectrum that is difficult to see visually. For this reason instruments with a CaK filter fitted tend to be used more for imaging applications. They pick out regions of glowing calcium atoms and reveal a region of the chromosphere that lies very close to the photosphere. Here the view is similar to that seen visually with the exception that the background granulation visible in white light is replaced by a network of faint glowing lines known as the chromospheric network.

The view here, apart from the more obvious chromospheric network, isn't dissimilar to that presented by a white-light filter, with the exception that features such as plage which appear clearly towards the edge of the CaK Sun are also clearly visible in the centre too. Bright prominences may also be seen using a CaK filter, although they tend not to be as dramatic as those seen through an H-alpha filter.

OUR FRIENDLY MOON

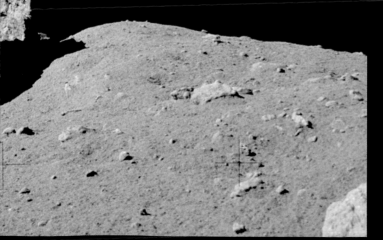

Apollo 17 astronaut Harrison "Jack" Schmitt at Tracy Rock (NASA).

CHAPTER 5

Just as the Sun dominates the day-time sky, the Moon is Queen of the Night. The main difference is that for part of the month the Moon is not always visible, though of course observing it is completely safe. It depends entirely on the Sun for its reflection and by direct observation you cannot damage your eyes. The Moon is regarded as the Earth's satellite, though in some ways it may be better to regard the Earth-Moon system as a double planet.

Despite its prominence in our sky, the Moon is not the largest natural satellite in the Solar System since three of Jupiter's moons and one of Saturn's are larger. The ratio of its mass to the mass of the Earth is only 1 to 81. This makes all the difference, and the Moon's low mass and hence low gravity has not made it possible for our satellite to retain a dense atmosphere or any surface water. This means that life there, at least of our kind, can be discounted. Whether it did support any kind of life must be regarded as dubious.

How was the Moon formed? There are several theories, none of which are satisfactory. The original idea is that Earth and Moon formed at the same time from the solar nebula (a cloud of dust and gas) and are linked. The main problem is that the Moon's overall density is less than that of Earth.

The second theory is that the Moon used to be an independent body that came close to Earth and was unable to break free. This has been seriously considered, but would require a very special set of circumstances.

Today, most authorities, not all, favour the giant impact theory. In this picture, the youthful Earth was struck by a body, possibly comparable in size to Mars, and the cores of Earth and impactor merged. Debris was spread around and eventually collected to form the Moon. There seems to be no insuperable problem with this theory other than the fact that the densities of Earth and Moon are very different. The density problem is solved if the Moon formed from the less dense outer parts of the Earth. In any case it is safe to say Earth and Moon are much the same age.

Look at the Moon with the naked eye and you will see bright and dark areas. Any small telescope will show the dark patches, or maria, the lunar seas; originally it was thought they were

[1] Full Moon, photographed with a DSLR camera attached to a 4-inch refractor.

[2] Full Moon and waning gibbous phases.

[3] Three waxing crescent Moon mosaics captured on consecutive nights starting on the right.

genuine seas while the light areas were land. The old names are still used and they are certainly romantic even though they bear no relation to reality. Mare Crisium, Sea of Crises; Oceanus Procellarum, Ocean of Storms; and so on.

It is quite possible the maria were once seas of lava, not water. Certainly the Moon must have been very active volcanically, but now the activity has ceased and the Moon is inert.

Men have been there – Neil Armstrong and Buzz Aldrin were the first, in 1969. They found a world in some ways similar to ours, and in other ways utterly different. It lacks an atmosphere and the sky is black in day time – by shielding your eyes and dark-adapting you can see the stars against the blackness.

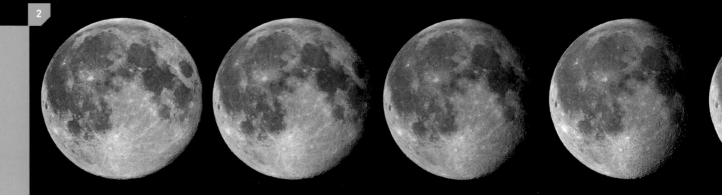

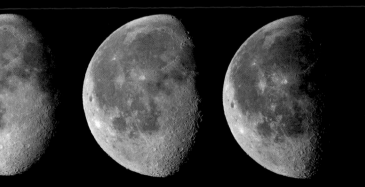

There can be no breath of wind nor weather. Since the surface is unprotected, the temperatures are extreme, ranging from very hot to very cold! The maria cover wide areas, and those on the Earth-turned side make up essentially a once-connected system, though some, notably Mare Crisium, are disconnected from the rest.

There is one very important point to be borne in mind. The Moon goes round the Earth, or to be more pedantic the barycentre (or centre of mass) of the Earth-Moon system, in 27.3 Earth days. This is also the period taken by the Moon to rotate on its axis. The result is that we always see the same side of the Moon, while the opposite side cannot be seen at all. Some find this hard to accept. We have even seen books saying the Moon does not rotate at all.

Imagine a model with the Earth in the centre, the Moon moving round it and the Sun off in the distance. As the angle between the Moon, Sun and Earth changes, the Moon shows phases or apparent changes of shape and these make what we call the lunation.

If the Moon lies virtually between the Sun and the Earth, its night side is turned towards the Earth. and we have a new Moon. In this position we cannot see it at all except when the alignment is exact and we see the glory of a total solar eclipse.

As the Moon moves along in its orbit, a little of its sunlit side turns in our direction. and the Moon appears as a crescent in the evening sky, thickening up until half the lighted side is turned towards us. Confusingly, this is named first quarter – the Moon has completed one quarter of its monthly journey. As the movement continues we see more and more of the sunlit side, and the Moon appears gibbous, meaning between half and full.

When the Moon is exactly opposite the Sun in the sky the entire day side is turned towards us and the Moon is full. The phases are repeated then in reverse order. Gibbous, half again (last quarter), crescent and back to new. Now suppose the Moon did not rotate. Over a month we would see the whole surface instead of only a fraction more than 50%.

The best example we can give is to imagine someone (the observer) sitting on a chair in the middle of the room and a companion walking round; as the walker moves, if he does not rotate he will face the same wall of the room the whole time and the observer will see the walker from every angle over the period of one rotation. This does not happen in the case of the Moon, which keeps the same face turned towards us all the

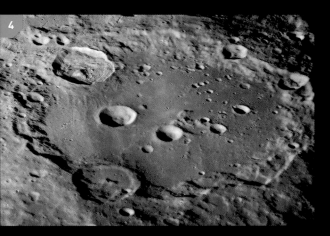

time, a situation known as captured or synchronous rotation. Almost all the major satellites of major planets behave in the same way. Tidal friction over the ages is responsible.

There is, however, one important qualification. The Moon's path around the Earth is not perfectly circular and this means its orbital speed varies slightly, quickest when nearest the Earth (perigee) and slowest when furthest away (apogee). This means that over one orbit the amount of spin becomes out of step and we can see alternately first round one limb then round the other. The Moon's orbit is also inclined by a small amount, allowing us to see a little beyond both poles at different times. This means that all in all we can examine a total of 59% of the entire surface though no more than 50% at any one time.

The areas on the edge of the Moon that appear and disappear as the Moon rocks and rolls its way across the sky are the

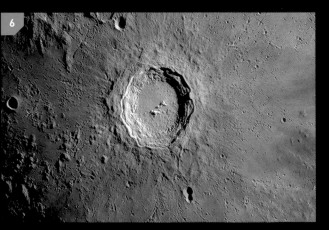

so-called libration zones. PM spent more than 30 years trying to compile charts using telescopes of various kinds, including the 15-inch in his own observatory. Now that spacecraft have been round the Moon and mapped the entire surface, he can look back and see how accurate his efforts were. On the whole they are pretty good.

Lunar Craters

Craters cover the entire Moon. Some are regular, with high walls and a central peak. Others have lower, flatter walls and no peak. Also the craters are of very different ages and many have been damaged by later impacts. For years there was a fierce argument about whether they were volcanic structures or formed by impact. PM "admits" he was on the wrong side of the argument, convinced they had volcanic origin, and only when the evidence for impact became overwhelming did he have to admit his mistake.

Note, too, that the craters are not shaped like steep-sided mineshafts. Seen in profile they are much more like saucers (not flying saucers, please note), some of them of immense size, well over a hundred miles in diameter. One of the most prominent on the Moon's nearside is Clavius, with a diameter of 144 miles and a string of craterlets on its floor. Basically it is circular but craters near the limb are often so foreshortened that they appear as ellipses and in some case it is not easy to

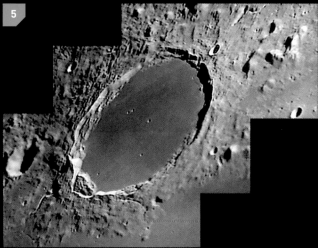

tell a crater from a ridge. It was this kind of thing that made mapping the libration areas such a big problem before space photography.

Just as an example, Copernicus has a diameter of 56 miles, high continuous walls and a lofty central mountain mass. It is also the centre of a system of bright streaks or rays. Tycho, 54 miles across, in the southern uplands, is of the same basic type but is considerably foreshortened with an even greater system of rays. They become evident under high lighting. Near full Moon, the rays from Tycho and Copernicus dominate the

[4] The floor of Clavius is littered with smaller craters.

[5] Crater Plato has a smooth floor pitted with craterlets.

[6] The impressive ray crater Copernicus with its terraced walls and central mountain peaks.

[7] The far side of the Moon photographed by the astronauts on board *Apollo 16*.

entire Moon. Tycho is often mistaken for the Moon's polar crater but this is not the case. It is some way from the Moon's south geographical pole.

Small craters are often bowl shaped and they too may have central peaks, though in no case does a central peak surpass the height of the highest mountains on the wall. Many craters have been so distorted by impacts that they are barely recognizable. For instance, Stadius, near Copernicus, once a noble formation, now almost classifies as a ghost.

Plato is another famous crater, in the north of the Moon, which appears as an ellipse even though it really is a circular crater 60 miles across. The walls are of modest height, and there is no central mountain and very little detail on the floor. It is one of the flattest parts of the entire Moon. It is also very dark grey so Plato is highly recognizable whenever it is in sunlight. An even darker-floored crater is Grimaldi on the Moon's west limb as seen from Earth. It is larger than Plato though less regular, and with a floor so dark it is impossible to overlook whenever it is in sunlight.

The far side of the Moon, invisible from Earth, is also mountainous and cratered, though there are differences in detail. There are mountains, seas and plenty of craters of all kinds. From this part of the Moon the Earth can never be seen because it always remains below the horizon.

It is amazing how quickly a lunar crater alters appearance according to the angle at which the Sun strikes it. A large walled plain is magnificent at sunrise, for example when shadow covers half of the floor, though under full illumination it may be hard to identify. Full moon is the worst time to begin observation, with sunlight coming straight down giving no shadows. The best views are obtained when the phase is crescent, half or gibbous.

The Moon may be changeless now, and certainly major upheavals belong to the distant past, but observers have reported indications of temporary localized glows or obscurations known as TLP (Transient Lunar Phenomena), a term for which PM was responsible.

The cause of these mild activities is a matter for debate, but disturbance of dust in the outer layer may be involved. TLP are not volcanic. Active vulcanism died out a long time ago.

It is fascinating to take a crater and draw it from night to night under different conditions. The differences are striking. There is tremendous scope here for the astronomical photographer and one does not necessarily need an expensive camera.

When seen in the night sky, the Moon appears bright and quite dominant compared to the fainter stars and planets. During the day, if the Moon happens to be above the horizon at the same time as the Sun, its appearance is more subdued, the surrounding blue sky swamping its light, making it harder to pick out.

The Moon can also occupy the transition zone between these two periods: dawn and dusk. All of these conditions provide opportunities for capturing the beauty of the Moon with a camera and as we'll see below, each requires a slightly different approach.

We'll start by discussing general lunar photography with a photographic digital camera. The brightness of the Moon at

[8] The subtle colour of the Moon brought out by image processing.

[9] Live view focusing.

[10] Earthshine captured using a DSLR camera.

[11] A planetary camera mosaic image of the crescent Moon.

night is often sufficient to allow your camera's automatic setting functions to work. Under these circumstances, set the camera to automatic, point it at the Moon and press the shutter button. The chances are that the photo you get back may not be that impressive to you.

Assuming it is exposed correctly and in focus, a typical camera lens delivers a Moon disc that appears tiny on the image frame.

Despite its dominance when visible in the night sky, the Moon's disc isn't actually that big, with an apparent diameter of 0.5 degrees. Put another way, hold your little finger up at arm's length and it'll easily cover the Moon's disc! The Moon's diameter in mm on a 35mm frame is given by dividing the focal length of the lens you're using by 109. If you're using a non-35mm-frame camera, as most are, the image will appear slightly bigger than this. So for a 200mm-focal-length lens, on a 35mm frame, the Moon's disc will appear just 2mm across!

A 300mm or larger telephoto lens will allow detail such as the lunar seas to be recorded but even at this length, the Moon's disc still appears rather small.

Things start to get interesting with larger telephoto lenses of 500–1000mm focal length; the longer the focal length, the easier it will be to make out more lunar features such as mountain ranges and craters. It's advisable to mount the camera and lens on a sturdy tripod to keep everything still during the exposure. A remote shutter release is also highly recommended.

Using a Telescope as a Lens

A telescope can be used as a lens for your camera. The way to achieve this depends on the nature of your camera. Fixed-lens cameras must be "connected" afocally, a term that basically describes the act of pointing the camera down the eyepiece. DSLR cameras can be connected at prime focus (the focal point of the primary mirror or objective lens), essentially direct-coupling the camera to the telescope. A description of these techniques is given in Chapter 2.

The settings you'll need to capture the Moon properly will depend on many factors, including the brightness of the background sky, the clarity of the sky and the focal length of the lens you're using. As a rule of thumb, if you're imaging using a photographic lens, set it a couple of stops back from being fully open (e.g. if a lens has a maximum aperture of f/2.8, knock it back to around f/5.6). Keep the ISO low initially too. Accurate focus is very important and can make or break the shot. For a telescope, focus as accurately as you can through the viewfinder. If your camera has a live focus capability which allows you to see what the sensor can see while focusing, use that. For a normal photographic auto-focus lens, turn the auto-focus on, point the camera at the Moon and half press the shutter button. If you're successful, the lens will have auto-focused on the Moon. When done, turn auto-focus off and be careful not to subsequently change the focus. Take your shot and review the result. If the Moon looks dark and underwhelming, increase the exposure. If necessary, increase the ISO as well but try to keep it below 400 to keep the noise levels down.

If the shot looks really bright, check for areas of pure white. If these exist, then the shot is over-exposed and needs to be done on a faster shutter speed and/or lower ISO. Many cameras can show over-exposed areas on their rear displays. If you're not sure what your model can do, check the user manual.

10

Deliberate Over-Exposure

When the Moon is a thin crescent, it's sometimes possible to see the dark, unlit portion glowing slightly against a background twilight sky, a phenomenon that when seen in an evening twilight sky is called "the old Moon in the young Moon's arms". The cause of the dim glow is the Earth. If you were to stand on the Moon's unlit surface at such times and look up in the sky, there you'd see the Earth at a near-full phase, shining down on you with reflected light from the Sun. This light would illuminate your surroundings in a similar manner to how the edge is taken off the darkness of night when the full Moon is visible on Earth. It's this gentle illumination that makes the unlit portion of the Moon appear to glow.

The effect is called "Earthshine" and is best caught on camera by using longer exposures that will cause the Moon's crescent to over-expose. For fixed-tripod shots, an exposure of a few seconds can, depending on the focal length of the lens you're using, introduce motion blur. Upping the ISO should allow you to capture Earthshine while keeping the exposure times to manageable levels.

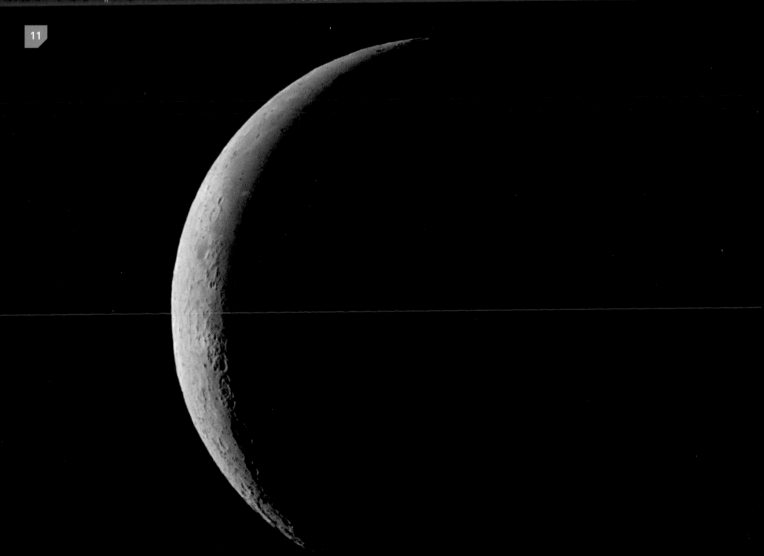

of the Moon like the surface of the full Moon, full of lunar detail.

Moon Colour

The Moon generally doesn't look colourful in most images. However, there is subtle colour to be seen on its surface, and amazingly, once the colours have been unlocked, they reveal some interesting scientific information about various features on the lunar surface.

A correctly exposed shot of the Moon's disc taken with a colour camera will contain this subtle information and it's possible to tease it out using a layer-based graphics editor. Load the image and then duplicate it as a second layer. Increase the saturation of the upper layer by a small amount, say 10–20%. Keep doing this until the colour is teased out of the image. If the coloured regions look noisy and garish, apply a Gaussian blur (a blurring widely used in graphics software) to this layer. Finally, set the blend mode of the layer to colour and merge back with the bottom layer. This last step brings back all of the luminance data – the detail – of the lunar surface, combining it with the colour that's just been teased out.

Going in Close

The techniques described above are generally useful for shots which cover the whole disc of the Moon or at least a large area of its surface. Increasing the focal length increases the magnification of the camera's view but will also reveal a limitation of still cameras, namely that imposed by the Earth's atmosphere.

As light passes through the atmosphere, its path gets refracted or bent by fast-moving blobs of air of differing temperatures and densities. The net result is a view of the Moon that appears to shimmer and shake. The stability of the view is described as "atmospheric seeing" and is a huge limitation to getting really high-magnification images of the Moon, filtered Sun and brighter planets. Fortunately though, there is a partial solution which can greatly reduce the effects of seeing – the high-frame-rate camera.

Unless the atmosphere is very unstable, it's normally possible to see brief periods of stability when looking through the eyepiece. The human brain is very good at ignoring the wobbly views and concentrating on the good stuff, but this is one area where a stills camera fails. Take a shot with a stills camera and unless you're very lucky indeed, the fine detail in your image will be distorted and blurred by the atmosphere.

The trick is to take lots of very short exposure still images in rapid succession in the hope that some of them record the less distorted views. Then, by pulling the good frames out of the pack, aligning them together and averaging the shots, the end result should be a relatively noise-and-distortion-free representation of the Moon.

The task of taking fast stills is handled by a high frame-rate-camera, something that can range from a PC webcam through to a high-end specialized planetary camera. These will typically capture still images at rates of between 5 and 120 uncompressed frames per second. It's common to store the resulting frame in a movie file format such as AVI.

The arduous process of pulling the good frames out of the collection is thankfully made significantly easier by a number of software applications, some of which are available to download and use completely free of charge. These programs can measure the quality of the still frames automatically, sifting them out for alignment (registration) and averaging (stacking). One popular freeware program to do this is called RegiStax.

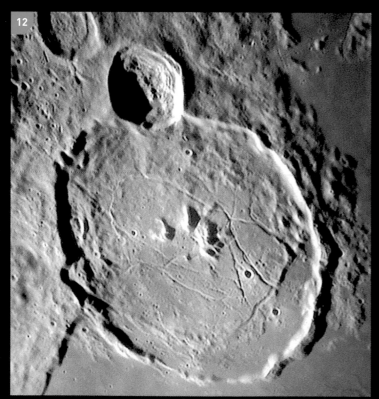

12

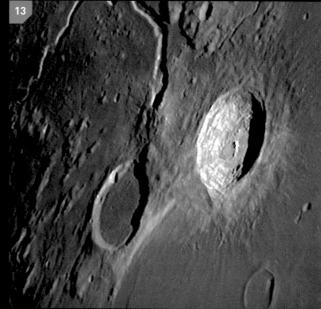

13

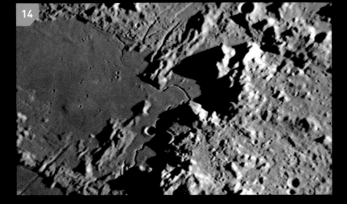

By using a high-frame-rate camera connected to a telescope, it's possible to create some amazingly sharp images of the Moon. Increasing the effective focal length of the scope by adding a Barlow lens makes it possible to magnify quite small regions of the lunar surface and pull out incredible detail. Barlows, remember, are optical amplifiers: optical systems which increase the effective focal length of your scope by the power of the Barlow. For optimum work, it's worth keeping the focal ratio, that's the effective focal length divided by the aperture of the scope using the same units, between f/25 and f/40, the upper range only being feasible if the seeing is really good to excellent.

If using a mono high-frame-rate camera, the seeing can be steadied further by fitting a red or infra-red filter over the nose-piece of the camera. Longer wavelengths (red light having a longer wavelength than blue) tend to fare better during their passage through the atmosphere, and using one of these filters can give the camera a steadier view. If the seeing is already very good, a green imaging filter can be used. The wavelength of green being shorter than red, this can produce even sharper end results.

For the best results, it's essential to take your telescope outside several hours before you begin your imaging run. This will allow the main optical tube time to cool, reducing the possibility of unwanted internal air currents. A 12V camping hairdryer attached to a suitable 12V battery source can be used to keep dew and moisture at bay from optical surfaces if this becomes a problem.

Lunar Mosaics

A typical high-frame-rate camera will have a relatively small imaging chip compared to a general photographic camera. For the fastest-rate cameras, image sizes of 640 x 480 aren't uncommon and, coupled to a telescope, the area imaged can be quite small. Larger shots can be created by combining smaller images together in what's known as a lunar mosaic. The easiest way to create a mosaic is to keep the camera settings fixed and methodically image sections of the lunar surface, ensuring you have large overlaps. The resulting captures must then be processed using a registration/stacking program as described above, ensuring, once again, that the settings used are consistent for all captures.

Using a layer-based graphics editor, the first image is loaded and defined as the reference. The canvas size of the image – the amount of working space around it – needs to be increased to accommodate the other sections. The second image is loaded as a new layer and nudged into position so that its overlapping features align with the reference image. A gentle 10% eraser tool can be used to remove the straight edges of the upper layer to assist the blend. This process is then repeated for the other sections until the mosaic is completed.

For those who wish to take an easier path, mosaic composition software can automate the process. An example of such software is the Microsoft Image Composite Editor (ICE) which, at the time of writing, is available to download and use free.

Summary

This may be a very rudimentary description of our friendly Moon but we hope it will suffice for the moment. We must always remember, it is the only world away from the Earth we have been able to visit. It would be perfectly possible to set up a permanent lunar base. The problem is with finance and political will. As long as mankind is intent on spending money on war rather than progress, the outlook appears rather gloomy! Meanwhile, the Moon awaits us and will not go away. In the future, the Solar System might have two inhabited worlds rather than only one.

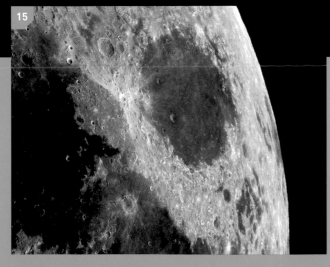

[12] Cracks cover the floor of crater Gassendi.

[13] The brightest crater on the earthward side of the Moon – crater Aristarchus.

[14] Hadley Rille imaged with a 14-inch SCT using a high-frame-rate camera fitted with a red imaging filter.

[15] Part of a massive mosaic built from 62 separate but overlapping images of the Moon taken through a 14-inch SCT using a high-frame-rate camera at prime focus. The circular sea visible in this shot is Mare Crisium.

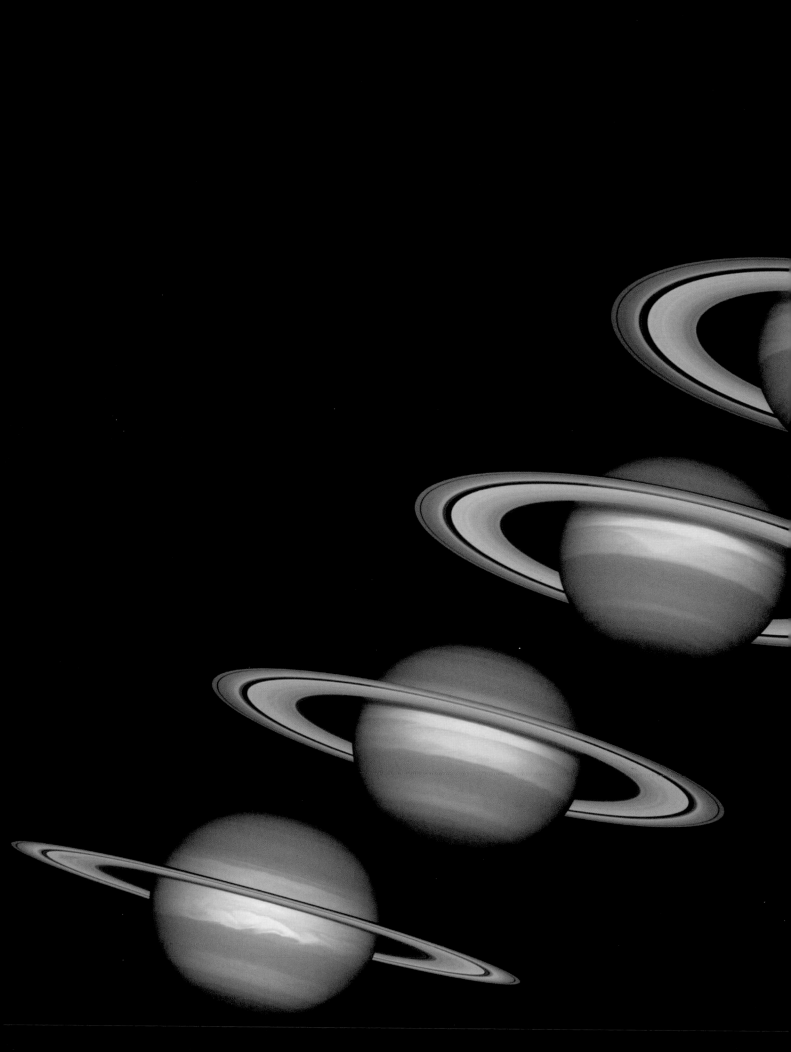

THE PLANETS

Saturn at different tilt angles imaged by the Hubble Space Telescope (WFPC2) between October 1996 and November 2000.

CHAPTER 6

The planets go round the Sun, rather than the Earth, and are very much further away than the Moon. The closet planet, Venus, is always at least a hundred times as far as the Moon. To observe the moons of Jupiter is an inspiring experience. To view and image Saturn and its rings is perhaps the most breathtaking experience in the whole of astronomy.

The Solar System is divided into two well-marked parts. First we have four small solid rocky planets: Mercury, Venus, Earth and Mars. Beyond Mars there is the wide gap filled by tens of thousands of dwarf worlds making up the main belt asteroids. Of these only one, Ceres, is as much as 500 miles across and only one, Vesta, is ever visible with the naked eye. The rest are much smaller and many are just pieces of rock. Just how they were formed is still a matter of debate; in all probability a planet is unable to form in this part of the Solar System because of the destructive gravitational force of giant Jupiter, more massive than the rest of the planets combined.

Beyond the asteroid belt we come to the four giant planets: two gas giants, Jupiter and Saturn, and two ice giants, Uranus and Neptune. All are much more massive than Earth and they are made up in a different way. We are also dealing with much greater distances. The further from the Sun, the larger the orbit and the slower the planet moves; Neptune, for example, takes 165 years to complete one journey round the Sun. All the giants have quick rotation periods, less than 10 hours in the case of Jupiter.

There is still discussion about the way in which the planets were formed, but certainly they condensed out of what we call the solar nebula, a cloud of dust and gas surrounding the youthful Sun. The fact that they are so different in mass means they evolved in different ways. Obviously, tremendous attention has been paid to the search for possible life. So far there has been no success at all, and we have to admit that at the moment the only life we know in the entire Universe is here on Earth.

Other stars have their own planetary systems. These must be dealt with separately, because they are so far away that we cannot see any detail.

All the Solar System planets are fascinating telescopic objects, and even a small instrument will give superb views. Moreover, the planets, unlike the Moon, are always changing and there is always something new to see. Neither does one know quite what to expect, and every planet can spring remarkable surprises on the unwary observer.

[1] Trails of Venus and Mercury over Portsmouth, England.

[2] Mercury seen from the *MESSENGER* spacecraft (NASA).

Mercury

Mercury is the innermost planet and the smallest. It is only 3000 miles in diameter, much less than the 7900 miles for the Earth, and its force of gravity is much lower. The escape velocity of a planet is the velocity needed to escape from the surface with no extra impetus. The value for Earth is 7 miles per second. The particles making up our atmosphere cannot move as fast as this, so we hold on to our atmosphere. Mercury's escape velocity is 1.5 miles per second, and if there once was a dense atmosphere there it has now leaked into space, leaving Mercury virtually airless. Without air there can be no water and no life of our kind. Mercury is unfriendly in other ways too. It is so near the Sun that during the day the hottest part of the surface is so torrid that life could not exist. At night, on the other hand, the temperature sinks very low because there is no atmosphere to blanket in the heat.

Because Mercury is lit by the Sun and has an orbit around the Sun that is inside Earth's, it shows phases or changes of shape; its orbital period is 88 days. With the naked eye Mercury is well seen either in the west after sunset or in the east before dawn. It is never seen against a really dark sky and most people have

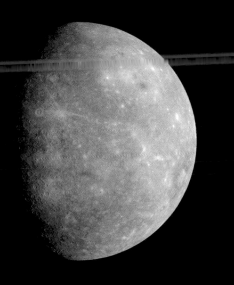

never seen it at all. When you do find it, it appears surprisingly bright – often brighter than every star in the sky. A small telescope will show patches, once regarded as seas, but we now know there has never been water there. There are mountains and craters, so in some ways Mercury is not unlike the Moon though here there are important differences in detail. Very large telescopes show much more of the surface, though by now we depend upon the results sent back by spacecraft that have flown past Mercury, and one probe, known as *MESSENGER*, that has been put into orbit around the planet.

It cannot be said that Mercury is a promising subject for the amateur observer; there is little to see apart from its phase. Occasionally the planet will transit across the face of the Sun, appearing as a small black disc against its surface. Transits are interesting to watch but not astronomically important.

Imaging Mercury

Mercury is a tricky planet to image because it never strays that far from the Sun. Evening appearances are better in the spring, while morning ones are better in the autumn, the planet appearing higher above the horizon at these times. Even then, catching sight of Mercury is limited to a few days before and after greatest elongation, when it appears furthest from the Sun.

Mercury's brightness fluctuates considerably when it is visible, because of its rapidly changing distance from us and the fact that it shows a phase (the changing illuminated portion of the planet that we see). Orbiting the Sun every 88 days, Mercury appears a bit like a cosmic moth, flying endlessly around the intensely bright Sun.

One of the best ways to capture Mercury with a camera is to use a standard or telephoto lens. Appearing as a dot on the final image, there's something very satisfying about grabbing a permanent record of this elusive little world.

The camera should ideally be mounted on a tripod to keep things nice and steady. If you're using a simple point-and-shoot camera, try to record the scene in fully automatic mode first. The chances are that the sky will record too brightly and Mercury won't be seen using this method, but it's worth a shot.

If auto doesn't work then you'll have to resort to manual settings. If you're not sure what your camera can offer here, the best advice is to look up "manual settings" in the camera's documentation. The main settings you'll need to get to grips with are focus, ISO, lens f-ratio and exposure.

Focus needs to be set to infinity. Here, our Moon can help if it's visible. Turn auto-focus on, point the camera at the Moon and half press the shutter button to focus on it. Once in focus, turn auto-focus off, and the camera should now be properly focused. If the Moon's not up, focus on a bright star or planet as accurately as you can. If your camera has a live view option, use that for greater accurary. The ISO setting you'll use will be determined by the brightness of the scene. As Mercury is typically located against a bright twilight background, a low ISO of 100–200 is probably the best starting point.

The lens f-ratio is the setting that controls how open the lens aperture is; in other words the control that dictates how much light is let into the camera. This should be set fairly wide (low f-ratio number) but it is best to avoid the absolute lowest setting as this can sometimes release aberrations inherent in some lenses.

For the exposure, start with a shutter speed of 1 second. Take a shot and view it on the camera's review screen. If the shot is too dark, increase the exposure time. Over a certain exposure length, Mercury will no longer appear as a dot but will instead appear to trail. This can be countered by increasing the ISO and decreasing the exposure.

It's very difficult to give absolute settings for different camera set-ups when attempting to shoot against a twilight sky. Cloud cover coupled with ever-changing light levels means that you typically have to adjust and adapt your settings to the conditions. Here the review screen is perfect because it can be used as a feedback device to let you know if your settings are off. If the image looks too dark, increase the exposure and/or the ISO setting. If the image looks too light, decrease these values.

Many cameras will let you know if the image contains areas of white indicating over-exposure that should be avoided. Methods may be different between camera models but a common way a camera shows this is to flash the areas of pure white in the review image. Again, if you're not sure whether your camera can do this, check its documentation. If you do have regions of pure white in your shots, reduce the exposure time and/or ISO setting until they disappear.

Using the techniques described will get you a shot of Mercury as a dot in either the morning or evening twilight sky. The presence of the Moon or even another planet nearby can add an additional level of interest to the composition. Alternatively, if you are using a fixed tripod to hold your camera, dropping the ISO sensitivity to its lowest value and closing the lens aperture to say f/11 or f/22 will reduce your camera's sensitivity enough to allow longer exposures. These will show Mercury as a trail in the sky rather than as a dot.

Another useful technique is to try to capture Mercury over the course of several days, at the same time, using the same lens settings and from the same location. If you then load each image into a layered graphics editor so that the horizons all line up, setting each upper layer to lighten mode will allow the previous position of Mercury to show through. This can then be used to reveal the apparent motion of Mercury in the sky.

It is possible to apply high-frame-rate imaging to Mercury but its poor placement in the morning sky prior to sunrise or evening sky post sunset tends to lead to a rather disturbed view of the planet. Advanced systems can locate Mercury in the day-time sky for a better-quality view but the dangers of the Sun being nearby should not be ignored.

[3] Venus rising, caught with a DSLR camera using a 28mm lens set at f/3.5. The camera was set to ISO 400 and the exposure was 14s.

[4] Venus, shot with a DSLR camera using a small aperture, shows diffraction spikes caused by the blades of the camera's aperture stop.

[5] Crescent Venus captured with a high-frame-rate camera fitted to a 14-inch telescope.

Venus

The next planet, Venus, is as unlike Mercury as it is possible to be. It is almost as large and massive as the Earth and, far from having no atmosphere, it has too much. Telescopically all one can see is the top of a layer of cloud. There was a time, only a few tens of years ago, when Venus was regarded as a possible abode for life, but later results have shown this cannot be so. The temperature is near to 1000° F, the atmospheric pressure is crushing, and the atmosphere is carbon dioxide with clouds of deadly sulphuric acid. Holidays on Venus are emphatically not to be recommended.

While Venus and Earth are near perfect twins in terms of size and mass, why are they so different? It may well be that when they were formed the Sun was much less luminous than it is now, so that Earth and Venus may have started to evolve along similar lines. But then over the ages the Sun became more powerful. Earth was out of the danger zone. Venus, at 67 million miles, was not. The temperature rocketed and the carbonates were driven out of the rocks, releasing the carbon dioxide we are familiar with today. Venus changed from a welcoming world into an immensely hostile one. Unmanned probes have been there and even landed on the surface, but whether manned flight will ever follow must be regarded as decidedly dubious. Certainly this will not happen for a very long time.

To the naked eye, Venus appears lovely, shining like an intensely bright lamp in the sky, so it is no wonder the ancients named it after the goddess of beauty. They had no way of knowing how hostile it was.

Imaging Venus

Venus, like Mercury, is an inferior planet, meaning that its orbit is smaller than that of the Earth. However, Venus's orbit is such that the planet can achieve a reasonable separation from the Sun when it's at elongation. Some elongations, it has to be said, are better placed than others.

Unlike Mercury, Venus always looks bright, due to its highly reflective cloud layer. Photographing Venus as a "dot", as described for Mercury, works well and is somewhat easier due to the planet's brilliance. Amazingly, Venus is also bright enough

to cast its own shadow and this too can be photographed using an exposure length of several minutes' with the camera set on high sensitivity. Both Mercury and Venus show phases. When on the closest part of their orbits to the Earth, both show a slender, delicate crescent. In the case of Venus, this can be quite large at almost an arc-minute (approximately 1/30th the apparent diameter of the Moon in the sky) across and is definitely a worthy target for any astrophotographer.

A rather ironic fact about Venus from an imager's point of view is that this most brilliant of planets is somewhat disappointing in terms of detail. Visually its disc doesn't give up its secrets easily and can appear rather bland.

The most successful methods for picking out subtle shaded detail within Venus's clouds is to use a greyscale high-frame-rate camera fitted with a selection of imaging filters. Most detail appears at the blue-UV end of the spectrum and the use of an imaging blue filter, Wratten 47 (violet) or a UV pass filter will reveal progressively more detail, assuming your imaging system is capable of showing it.

One problem when working with a UV filter is that the image is often incredibly dim, making it hard to achieve a decent focus. Another issue is that modern optical coatings can interfere with short wavelengths, reducing the amount of detail transmitted. This also applies to Barlow lenses used to increase image scale and has led a number of amateurs to switch to using fused silica Barlows when imaging Venus to limit UV loss.

Neither Mercury nor Venus has satellites. Both are solitary wanderers in space.

Next in line our own Earth is the only planet suited for the development of intelligent life. (Whether it actually appeared here is open to debate!) Beyond comes Mars, the planet we have observed and explored in most detail through the eyes of NASA's surface rovers.

Mars

Mars's year is 687 Earth days, and it has a rotation period of 37 minutes longer than our own. To us Mars has always held special interest, being the only world where our type of life could exist; a few tens of years ago it was believed Mars was the centre for established life, with canals visible through our largest telescopes. The canals do not exist; they were tricks of the eye, and Mars today is a world where we could not survive in the open.

The planet is a rocky desert and though it must once have had wide seas, the water has dried up and today the surface of Mars is completely arid.

Small telescopes will show the main surface details, the red deserts, the dark areas and the ice caps at the poles that wax and wane with the Martian seasons. There is no doubt the caps are made of water ice, but when they shrink in the Martian summer they do not melt but sublime – change directly from a solid into the gaseous state. Rainfall has been unknown for

[6] Mars Rover *Spirit* on the surface of the Red Planet, looking towards Husband Hill (NASA/JPL/Cornell).

[7], [8] & **[9]** Images of Mars captured using the imaging technique described here. A 14-inch SCT telescope was used throughout.

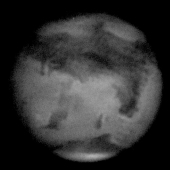

millions of years. The atmosphere appears to consist mainly of carbon dioxide with not much free oxygen.

Many spacecraft have now been sent to Mars. Some have been put into orbit and mapped the entire surface. Others have crash-landed there, and two, the US *Spirit* and *Opportunity*, were brought gently down and roved around the surface sending back data. At the time of writing *Opportunity* is still moving around and sending back all kinds of fascinating information. The second probe, *Spirit*, became stuck in a drift and finally lost power in 2010. But it did its job well, even better than expected. The latest probe, *Curiosity*, landed in 2012 and is the most complicated vehicle ever to be sent to Mars.

Winds on Mars are quite strong, though in that thin atmosphere they have very little actual force. It was once thought that the atmospheric pressure at ground level was 85 millibars – the actual ground pressure is below 10 millibars everywhere, corresponding to what would be a laboratory vacuum on Earth.

There are craters due to impact, not unlike those of the Moon. There are also high mountains and huge volcanoes that must once have been very active but are now, we believe, extinct. The highest mountain, Olympus Mons, towers to a height of 25 thousand feet above the ground below, and is crowned by a huge summit caldera. The orbiting probes have sent back pictures which are very impressive. No doubt it will be climbed one day – not too much of a problem because the slope to the summit is very gentle. A spacesuit would not be an insuperable problem.

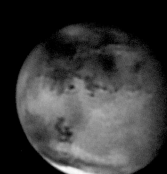

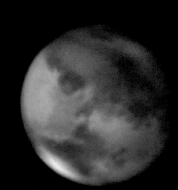

When will people reach Mars? One problem is radiation. In the journey there, which will take several months, the travellers will be in their craft, unprotected, bombarded by all kinds of radiation. No one could stand up to that, as far as we know. No doubt there is a cure for this and some kind of lightweight radiation screen will be devised. So far we have no idea how that can be done. Until we solve the radiation problem, we believe Mars is out of our reach.

Mars has two moons, Phobos and Deimos, discovered by the US astronomer Asaph Hall in 1887. Both are very small, Phobos less than 20 miles in diameter and Deimos less than 10. Almost certainly they are captured asteroids from the main belt moving round the Sun beyond the path of Mars.

The Martian day is just over half an hour longer than ours, but the Martian year is equal to 687 Earth days, equating to 686 Mars days, or sols. The seasons are the same basic type as ours though of course much longer.

We have one idea at this point. Phobos and Deimos make magnificent natural space stations. So small they have almost no gravitational pull, so landing and takeoff would be easy. The first journey to Mars might well be via a Deimos base, with a short hop down to the surface of Mars. Time will tell.

Imaging Mars

Mars is a fascinating world. The most obvious features are the contrasting light sandy regions representing the deserts of Mars and the darker, exposed rocky regions. As these are primarily seen by virtue of the light they reflect, they are known as albedo features. It is possible to image more detailed relief features such as the larger craters, volcanoes and the Martian equivalent of the Grand Canyon – the Vallis Marineris – but this requires a big scope and extremely good atmospheric stability.

The best results with Mars are to be achieved by using a high-frame-rate camera. Both colour and mono cameras can achieve excellent results, with the latter being better suited to high-resolution imaging. A set of imaging filters can be used to capture a red, green and blue (R, G and B) image set, which can then be recombined in a graphics editor to produce a full-colour version of the planet.

Telescope optics should be properly adjusted (collimated), and the telescope should be given sufficient time to cool outside so that it is in thermal equilibrium with its surroundings. An hour or two outside will achieve this for most telescopes, although large ones may need longer.

Point the telescope at Mars and insert the camera at prime focus (no Barlow lenses). Focus and assess the atmospheric stability (seeing). If the seeing looks good and the planet steady, consider inserting a Barlow lens to up the image scale. Aim for a typical workable focal ratio for Mars of f/25–f/40, achieved by multiplying the focal ratio of your scope by the power of any Barlow used. For example, an f/10 instrument used with a 2x Barlow will effectively work as an f/20 scope. Only consider going to the upper end of the scale if the seeing is good.

Record a movie sequence using R, G and B imaging filters in turn. For high-resolution work with focal lengths greater than 10 metres, do not take more than 1–2 minutes per channel, otherwise the planet's rotation will cause motion blur. Interestingly, as Mars shows little useful detail through a G filter, it is possible to use a technique for generating a synthetic G by averaging an R and B result together.

This is done by loading both the R and B results into a layer-based graphics editor and setting the transparency of the upper layer to 50% before merging them together to form the synthetic G.

Common notation shows the use of synthetic channels in

brackets, for example R(G)B. The advantage to this technique is that the overall capture time is shorter, allowing you to complete a colour image in a third less time than it would take for a full RGB set.

The individual RGB or RB captures need to be processed using a registration-stacking program. The most commonly used applications for this stage are the freeware programs Registax and AVIStack.

Once you have all three channels – RGB or R(G)B – use a graphics editor to combine the channels. Each channel image must be loaded into the individual colour channel of a blank RGB image, R being loaded into the red channel, G or (G) into the green and B into the blue. The G/(G) and B channels must also be aligned with the R.

Once done, the final result will be a full-colour rendition of Mars. Typically, this will need to be massaged to a final end result using the editing program's levels, curves, contrast and brightness adjustment tools. A gentle sharpening to the end result can be applied if appropriate to make detail clearer to see.

Most of the planetary detail is recorded in the R channel, with the B being particularly good at picking out any clouds present in the Martian atmosphere. The use of an infra-red (IR) pass filter such as the popular 742nanometer (nm) "IR pro planetary filter" will produce a dimmer image to work with, but often higher-contrast details.

A sharp R or IR image can be re-introduced to the main RGB result by layering it over the main colour image and setting its blend mode to luminance. Applied like this, the sharper R

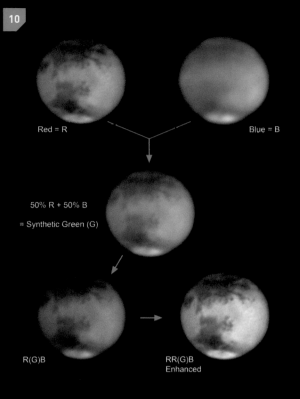

Red = R

Blue = B

50% R + 50% B
= Synthetic Green (G)

R(G)B

RR(G)B
Enhanced

or IR image supplies the detail to the image while the RGB result supplies the colour information. Such images are normally labelled R-RGB or IR-R(G)B to indicate the use of a luminance channel.

Mars has a somewhat frustrating rotation, taking almost 40 minutes longer than an Earth day to rotate once on its axis. This means that if you image at the same time of night when Mars is close to its highest point in the sky, it will take Mars another 40

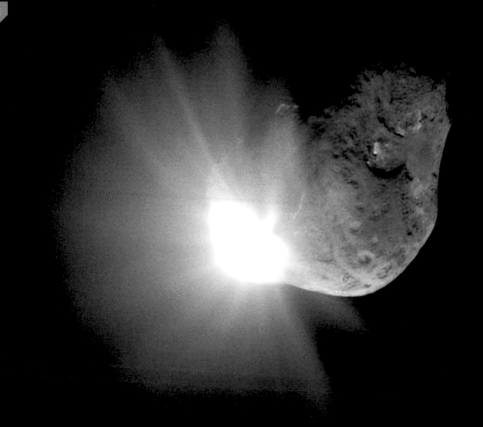

[10] Diagram shows the process of combining a red [R] filtered image with a blue [B] one, using half from each to produce a synthetic green [(G)]. These three components are then combined to produce an R(G) B image as shown in the lower left. Finally, adding the R back in as a luminance layer and tweaking the result gives a very detailed image of Mars.

[11] Comet Tempel 1 impacted by the probe *Deep Impact*.

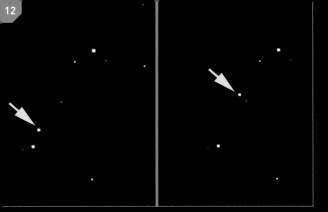

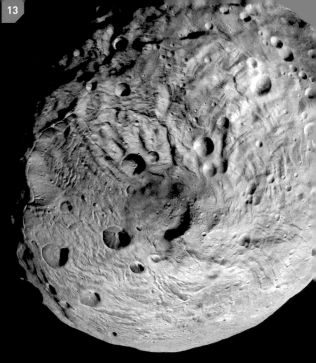

12] Asteroid imaged using the technique described below.

[13] Asteroid Vesta, seen from the *Dawn* probe (NASA/JPL/Caltech).

minutes to rotate back to the same point it presented the night before. Over subsequent nights at the same time, you slowly get to see a small amount of new surface detail before the planet rotates into the same position it started from the night before.

Asteroid Belt

Beyond Mars lie the main belt asteroids, tens of thousands of them. There is a book by a well-known writer called G.F.D. Chambers, the first astronomical book PM read at age seven. In this book Chambers refers to the asteroids, and says, "They are of no real interest to the amateur observer, and they are in fact of no interest to anybody!" He could not have been more wrong. We can learn a great deal from the asteroids. In 2011 the spacecraft *Dawn* entered orbit around Vesta, the third-largest asteroid, to conduct a survey before heading for Ceres in 2012, now ranked as a dwarf planet.

All the large asteroids keep well clear of the Earth but smaller ones do not and some have been known to pass between the Earth and the Moon. There is always a danger we will be hit by one, with catastrophic results. There is no difference between a large meteorite and a small asteroid. Amateur astronomers are experts in photographing the Near Earth Asteroids so their orbits can be worked out and we know when to expect them.

We have to admit that if we saw a mile-wide asteroid approaching us we could do nothing about it. Once PM had a letter from a lady asking what to do in the event of an imminent impact from an asteroid. He replied, "Repeat after me ... Our Father..." However, the chances of this happening in our time are very low.

The brightest asteroid, though not the largest, is Vesta, roughly 300 miles in diameter. In 2011 the probe *Dawn* went into a path around it and returned photos showing a rough cratered surface with one huge impact crater near the pole. Having made a survey, the *Dawn* probe set off for an encounter with Ceres. They will take some time to analyse, but we can expect significant results from that to appear any time now.

Asteroids are the flotsam and jetsam of the part of the Solar System in which the main planets orbit. They mostly inhabit the region between the orbits of Jupiter and Mars and range in size from about 600 miles in diameter down to a few tens of feet across. The largest one, Ceres, is no longer classified as an asteroid as such, having been promoted to the class of dwarf planet.

Imaging Asteroids

Visually, an asteroid looks no different to a star and this makes it harder to identify. The way to confirm that you've seen an asteroid is to record the field in which you think it lies and do the same on a subsequent night. If you compare the recordings and one of the "stars" has moved, that is probably the asteroid.

Sketching the field is one way to do this, although a more up-to-date method would be to image. Here, any camera that can take a photograph of a star field can be used. It's normal to couple the camera to a telescope in order to record a field of view with a good image scale and with sufficient light depth to record the asteroid convincingly. If this is done over several nights the images can be brought together in an animation. For this to work, the images need to be taken with the same equipment and cover the same star field. Post capture, each image needs to be aligned with respect to its stars before being added to the animation. Running the sequence in a loop will "blink" between the frames, revealing the motion of anything that is moving.

For a typical asteroid, a separation of one, two, three or even more days between shots is normally fine, but there is a class of asteroid that appear to move so fast across the sky that they will show motion after hours or, in extreme cases, minutes. These are the so-called Near Earth Asteroids or NEAs for short. As their name suggests, these are bodies that have

15

[14] & [15] Jupiter imaged using a mono high-frame-rate camera and the RGB technique described.

orbits that take them relatively close to Earth, sometimes even closer than the Moon. Consequently, these bodies can appear to move quite rapidly across the sky, the apparent speed reaching a maximum as the asteroid's distance to Earth reaches a minimum.

Catching an NEA on camera requires skill and timing, but in practice the procedure to follow is identical to that described above for "normal" asteroids. Predictions of passes can be obtained from websites such as that provided by the Jet Propulsion Laboratory at http://neo.jpl.nasa.gov/neo/. Accurate positions can be obtained by entering the object's reference details along with your own position etc. into the JPL Horizons ephemeris generator at http://ssd.jpl.nasa.gov/?horizons.

Once you know the star field where the NEA is going to pass

through, it's a case of setting up your camera and waiting for the appropriate time to catch it. It is important to check the magnitude of the body and that your imaging set-up can record stars at least down to that level, preferably a bit dimmer.

Important work can be done in this field by measuring the position and brightness of the object. Accurate brightness determination over short exposures can be used to measure the rotation period of the asteroid.

We now take the next major step away from our home planet Earth, and head into what is known as the Outer Solar System – where rocky planets give way to the giants – Jupiter, Saturn, Uranus and Neptune.

[16] A DSLR shot of Jupiter and Uranus taken through a 5-inch short-focal-length reflector. Note the Galilean moons close to Jupiter. Uranus is the bright dot above and left of centre.

[17] Amateur astronomer Damian Peach's prize-winning photograph of Jupiter, Io (left) and Ganymede. The image was taken through Damian's 14-inch SCT telescope from Barbados.

Jupiter

Out this far we find a very different side to the Solar System. When the outer planets formed, conditions were much cooler than nearer to the Sun, so these planets evolved very differently. The giant planet Jupiter has a core made up of ice and silicates surrounded by a dense atmosphere containing hydrogen, helium and other gases. A telescope shows Jupiter as a yellowish flattened disc. Although the orbital period is nearly 12 years, the rotation period is less than 10 hours and this rapid spin causes the equator to bulge out. Normally the disc shows streaks known as cloud belts. The most famous feature is the Great Red Spot, an anti-cyclonic storm with a surface area greater than the surface of the Earth. The cause of the red colour is still a matter for debate. It is out of view for half of Jupiter's day on the far side of the planet.

The joy of Jupiter from the observer's point of view is that it is always changing and one never knows what it will do next. Normally there are several belts, in particular the two equatorial belts, one either side of the planet's equator. In 2010 the south equatorial belt suddenly vanished. It might well have been snatched away like the "Hunters of the Snark" and Jupiter observers were baffled. It has since reappeared but is not as regular as it was and it will be some time before it reverts to normal. The disappearance is thought to be due to the high atmospheric clouds which temporarily obscured the belt.

Jupiter is the ideal subject for the visual observer and also for the photographer.

Imaging Jupiter

Jupiter is a wonderful planet to image because it presents a large disc, and is very bright and very detailed. It also rotates quickly, completing one rotation on its axis in less than 10 hours. Consequently, over the course of a night, you can get to see quite a lot of the planet as features rotate into and out of view.

The fast rotation also presents some interesting problems for high-resolution imagers because captures that take too long will blur detail. The maximum imaging time for Jupiter should ideally be kept under a minute per channel for RGB images at moderate focal lengths – 3 minutes maximum if using a colour camera. For long focal lengths greater than 10m, this should be reduced to 40s per RGB channel or 2 minutes if using a colour camera.

High-frame-rate cameras, with their ability to overcome some of the limitations imposed by seeing conditions, are the best instruments to use for Jupiter, with a bias once again going towards mono cameras using RGB imaging filters. The techniques to employ are similar to those listed for Mars and, once again, it is possible to utilize the synthetic green channel technique,

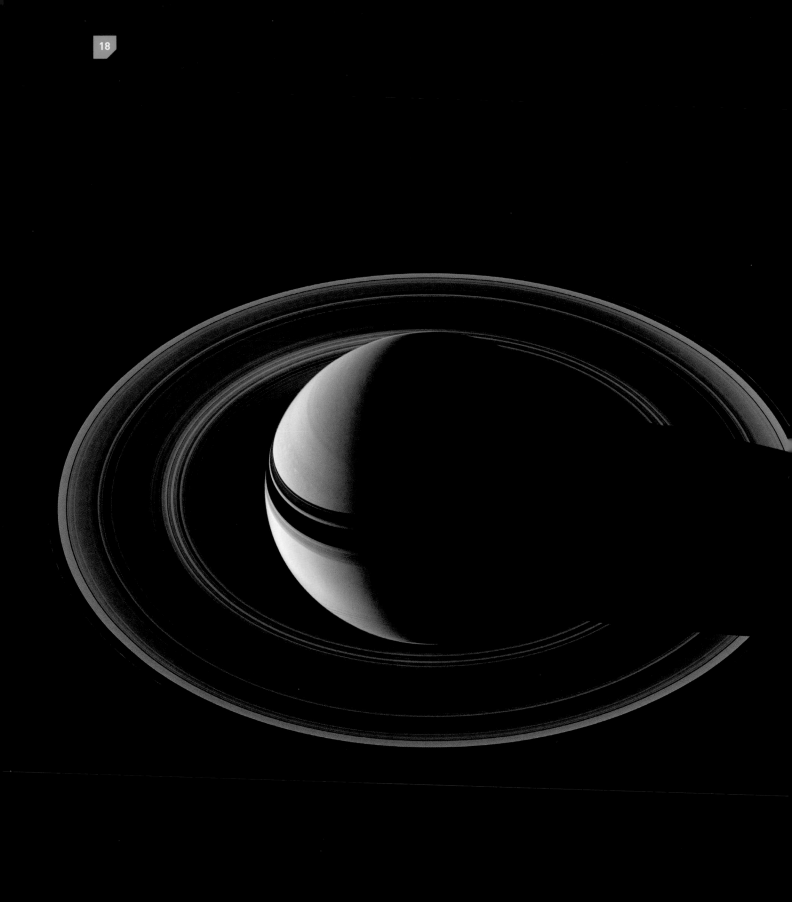

conveniently cutting RGB imaging time down by a third.

High-frame-rate colour cameras can also work well with Jupiter, allowing you to capture everything in a single and, more importantly, shorter sequence. Here, atmospheric dispersion – an effect that makes the colour from Jupiter spread out as if the planet's light is passing through a prism – plays an important part in trying to ruin the image. The effect is worse the lower a planet appears in your sky, as the amount of atmosphere its light has to pass through gets progressively thicker closer to the horizon. This is where a device known as an atmospheric dispersion corrector (ADC) can help. An ADC counters the atmospheric dispersion by introducing prisms of its own to recombine the colours. As dispersion varies with altitude, one downside of using an ADC is that it needs to be constantly adjusted throughout a typical imaging session for the best results.

As Jupiter is so bright, imaging it isn't something limited to the realm of the high-resolution specialist; some pleasing results can be achieved with more modest equipment too. A simple camera attached to a tripod will normally record Jupiter as a dot in the sky as long as the camera's sensitivity is set reasonably high. Many camera phones can capture the planet as a dot too.

For cameras with a zoom or telephoto lens, it is possible for a short exposure of the planet to show both the planet's disc and the Galilean moons – assuming they are visible at the time of the shot of course.

Jupiter's brilliance also sets it up as a good target for afocal photography, the technique of pointing a camera with a lens fitted down the eyepiece of a telescope. Although such images will normally appear small and slightly blurred, they can show the main belts of Jupiter as well as the four Galilean moons quite well. Some quite impressive shots of Jupiter with mobile-phone cameras have been taken using this technique.

Jupiter's Moons

Jupiter has a whole family of satellites, over 60 in all. Most are small and no doubt captured asteroids. Of the four large moons, Io is slightly larger than our Moon, Europa slightly smaller, and Ganymede and Callisto considerably larger; Ganymede is actually larger than the planet Mercury. All four are visible via any small telescope and they may pass behind Jupiter and be occulted, or eclipsed, or in transit across the planet's disc.

They are by no means all alike. Io is the most volcanic world we know, with violent eruptions going on all the time. It is right in the middle of the dangerous radiation zones surrounding Jupiter and we are not likely to want to visit it!

Europa has a smooth icy surface and probably an ocean of liquid water underneath. There might even be life present in these oceans, perhaps similar to the life that appears around hydrothermal vents deep in the darkest parts of Earth's oceans. Ganymede and Callisto are mountainous and cratered. All the other moons are much smaller

Imaging Jupiter's Galilean Moons

Jupiter has an ever-growing family of moons (at the time of writing the current count stands at 64). However, only four of these moons are easily seen, the rest being quite faint with many in inclined orbits so that they appear away from Jupiter's equatorial plane.

The four so-called Galilean moons, Io, Europa, Ganymede and Callisto, are bright enough that they can be seen through a steady pair of binoculars, and make excellent targets to photograph. Their brightness is such that even modest equipment can pick them up. As a bonus, as the four moons orbit close to the equatorial plane of the planet and Jupiter's tilt relative to the Sun is quite small at just 3.1 degrees, it is also possible to see a whole host of interesting interactions between them and Jupiter's disc.

These interactions include transits of the moons in front of Jupiter, occultations as they are hidden behind the planet, and eclipses as they enter Jupiter's shadow. The moons can also cast their own shadows on Jupiter's disc and these are quite easy to observe through even small scopes. On rare occasions when the tilt angle of Jupiter is low, it's also possible to see moons passing in front of, or casting shadows over, one another.

For the sharpest moon overview images, a high-frame-rate camera will produce the best results, but the telescope requirements here are quite modest. A 3-inch refractor on a driven equatorial mount is capable of producing a lovely image of Jupiter and its family of Galilean moons. The technique which works best here is to produce two images – one optimized for the planet's disc, the other set to record the moons themselves. The moons will appear indistinct or even absent in the first image while Jupiter will appear over-exposed in the second.

Once you have both captures, they can be combined in a layer-based graphics program such as Photoshop (commercial), Paint Shop Pro (commercial) or GIMP (freeware). There are many ways to do this but one is to layer the over-exposed Jupiter image on top of the surface-detail-optimized one. Align both images, then carefully paint out the disc of Jupiter in the over-exposed image using black. If the glare from the planet has created a bright halo, this too can be removed by over-painting if required. Once done, simply set the upper layer's blend mode to lighten and the bright moons will be joined by the planet's disc from the layer below.

Next on our list, moving outward from the Sun, we come to what in our view is not only the loveliest of all the planets, but the loveliest thing in the entire sky.

[18] The magnificent planet Saturn as captured by PL using his 14-inch Schmidt-Cassegrain Telescope fitted with a high-frame-rate mono imaging camera. A set of RGB imaging filters were used to capture three separate colour images which were then combined using Photoshop to produce the final image you see here.

Saturn

In itself, Saturn's not too unlike Jupiter. It is smaller and its inner temperature is lower. It certainly qualifies as a gas giant with a silicate core containing rocky particles and an immense surround of gas – mainly hydrogen with a good deal of helium. There is no solid surface to land on, and it is not easy to decide where the atmosphere ends and the body begins.

Several spacecraft have visited – *Pioneer 11* in 1979, then two *Voyager*s in 1980 and 1981, and the present *Cassini* probe, named after one of the best early observers of the planet, Giovanni Cassini (1625–1712).

Cassini went into orbit around the planet, and at the time of writing, early 2012, it has already lasted for some years and there is as yet no end in sight to its activity. Certainly it has increased our knowledge of Saturn by a factor of two.

The globe of Saturn shows belts similar to Jupiter but less prominent. There are occasional spots. In 1933 one was discovered by W.T. Hay, remembered as Will Hay, the stage and screen comedian. This was a white spot that PM remembers seeing very clearly in a 3-inch refractor. Other white spots have been seen since and may be commoner than we originally thought.

But of course the glory of Saturn lies in its rings. The other giant planets also have rings, but they are dark whereas Saturn's rings are icy and bright. They are in fact made up of particles of ordinary water ice, all spinning round the planet in the manner of dwarf moons. The individual particles are so small and remote that the rings have the appearance of being solid, but a moment's consideration shows this to be untenable.

No solid ring could form and in any case a ring of that kind would promptly be torn to pieces by Saturn's powerful pull of gravity. The idea that the rings are made of icy particles came from James Clerk Maxwell in the 1870s, and was confirmed by measurements showing the inner rings move round the planet more quickly than the outer rings. In fact they follow Kepler's laws; the closer to the main planet, the quicker they move. (Kepler was the 17th-century Germanastronomer who drew up the three famous laws of planetary motion – see the Glossary).

This is exactly how a ring made up of discrete particles would be expected to behave, but the rings are strange in another way. The overall diameter of the system, from one ring tip to the other, is 169,000 miles but the thickness is less than a mile, so when the rings are almost edgewise on they appear only as a thin line of light and small telescopes will not show them at all for several weeks. With a 12.5-inch reflector PM has followed them through the edgewise presentation, though not easily.

The ring system is complex. There are three main rings – two bright and one semi-transparent. The two bright rings are labelled A and B, separated by a gap named the Cassini division. B is closer in. Closer in still we come to ring C, the Crepe or Dusky ring, discovered in the middle of the 20th century. It is by no means a difficult object, but is semi-transparent and easily overlooked. One curious fact associated with ring B is the appearance of strange dark spokes that move around the planet and persist for some time before dissipating. They are due to particles elevated away from the main ring plane possibly by electrostatic forces, and are still something of a mystery. They are confined to ring B; A does not show them.

19

[19] Saturn seen from the Cassini probe (NASA).

[20] A crescent Enceladus appears with Saturn's rings in this Cassini spacecraft view of the moon. The famed jets of water ice emanating from the south polar region of the moon are faintly visible as a small white blur below the dark south pole. (NASA/JPL-Caltech/Space Science Institute.)

[21] Tethys and Titan seen from Cassini (NASA)19. Saturn seen from the Cassini probe (NASA).

RRGB

R   *G*   *B*

[22] Separate Red, Blue and Green (RGB) images taken with a high-frame-rate camera and combined with a graphics package.

[23] Saturn and its moons. From right to left: Titan, Dione, Mimas, Enceladus, Tethys, Rhea.

[24] Saturn showing the northern hemisphere storm in March 2011.

[25] Mimas, showing its largest crater, Herschel, at upper right (NASA).

[26] Uranus photographed through a 14-inch reflector, combined RGB images.

These are the rings seen through a modest telescope, but there are others that are much more elusive and beyond the range of the average amateur. Some move at distances further out than the main rings, and although the entire ring system is surprisingly extensive it is also very rarefied. Although Saturn is a giant world, it is also true that its overall density is less than that of water.

Imaging Saturn

The imaging techniques required for Saturn are similar to those described for Mars and Jupiter although it's not common to synthesize colour channels for this world. Capture times using high-frame-rate planetary cameras should be kept to less than 2 minutes per channel for RGB images and 4–5 minutes for full-colour captures. The biggest hurdle when imaging Saturn is the planet's brightness or rather lack of it. It can be a struggle

Saturn was the first thing PM saw through a reasonably powerful telescope, and he could not keep back a shout of wonder. This is the general reaction when people see Saturn for the first time. It looks somehow unreal, as though it is a model put there in the sky especially for our benefit.

The rings move in the same plane as Saturn's equator, and this means that over a Saturnian year we see them at various tilt angles. Saturn takes nearly 30 years to complete one journey around the Sun and for most of the time the rings are so tilted we can see them in detail. Drawings can be very instructive, though by no means easy to make, and once again the astronomical photographer has a major part to play.

Saturn is the most beautiful of the planets due to its stunning and very photogenic ring system, but its disc doesn't typically show a lot of detail in comparison with its nearer neighbour, Jupiter. Belts and zones can generally be seen, as well as bright patches which represent storms in the planet's atmosphere.

getting a signal strong enough on chip to do the planet justice.

The sharpest features on Saturn are where the planet's disc appears to interact with the rings and these are regions which are very prone to artefacts during the processing of high-frame-rate captures. Here, the seeing plays an important role too, certain types of fast-moving seeing appearing to "drag" some of these features to the side, making them look indistinct and blurred. The "gap" between the two brightest rings, A and B, known as the Cassini division, is especially prone to this effect.

If the seeing is very good, one difficult-to-image target, highly sought after by Saturnian imagers, is the 200-mile-wide Encke Division or Gap. This lies close to the outer edge of the A ring and requires a large scope and good, still atmospheric conditions to capture well. As was the case with Jupiter, low-power afocal shots can produce results, although the lower light levels mean you'll need to use longer exposures or higher (and noisier) ISO settings.

Saturn's Moons

Next let us turn to Saturn's family of satellites, which is of special interest. At the end of the 20th century, nine Saturnian satellites were known. Many have been added since though most are small – probably captured asteroids. Of the classic satellites the largest by far is Titan, discovered in 1665 by the Dutch astronomer Christiaan Huygens. Titan is of planetary size and is larger than Mercury though less massive; of all the satellites only Ganymede in Jupiter's family is larger. However, Titan is unique in possessing a dense atmosphere made up chiefly of nitrogen.

There is a sort of curious fact here. Retention of an atmosphere depends upon two things: first of all the escape velocity of the central body and secondly the temperature. Our Moon has an escape velocity of 1.5 miles per second, and is unable to retain a substantial atmosphere. On Titan, the escape velocity is much the same but the temperature is much lower, being so much further from the Sun. The higher the temperature, the quicker the various atoms and molecules move around, so that those in Titan's atmosphere are more leisurely than those of the Moon. This means Titan has no problem in hanging on to a thick atmosphere. None of the four large Jovian satellites has had equal success.

We know a great deal about Titan thanks to *Cassini*, which carried a much smaller probe, *Huygens*, which landed gently on the surface of the satellite. This is one of the most amazing feats of space research so far: bringing a probe down gently on the surface of a satellite nearly 900 million miles away. After landing, *Huygens* was able to go on transmitting data for some hours, and the results were fascinating. The landing site was "mushy", but there were lakes and seas within range, not of water but of methane. Titan is the only world apart from Earth where we have definitely detected liquid oceans. However, going for a bath in them is not to be recommended because they are, to put it mildly, distinctly chilly!

The rest of Titan has been well mapped by *Cassini* and been found to be a strange place. The methane clouds in the sky are never clear, so there is probably a constant methane drizzle, making Titan a rather gloomy world. There is no such thing as a sunny day there.

Fascinating though Titan is, it is surpassed by one of the much smaller satellites, Enceladus.

The two classic satellites closer to Saturn, Mimas and Enceladus, were discovered in 1789 by William Herschel with the aid of his new large telescope. Both are less than 400 miles across. Mimas is icy and inert, but it's Enceladus that is the real problem. Close-range views have shown there is a very tenuous atmosphere, and that vents near the poles are streaming out water. In fact it has water geysers. This means there must be an underground sea to provide the material. Enceladus, less than 400 miles across, is in theory far too small to show any trace of activity, let alone constant fountains. No one has yet come up

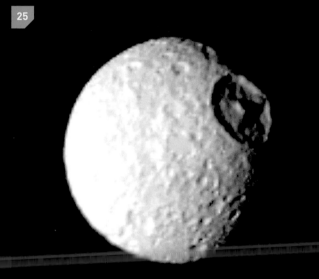

with a satisfactory answer. The fountains ought not to be there, but they are. It is doubtful if we will learn much more before we have new probes, possibly manned. A very long-term project indeed.

Beyond Enceladus there are three more icy satellites – Rhea, Dione and Tethys, all discovered by Giovanni Cassini. Beyond the orbit of Titan is a small irregular moon, Hyperion, a severe test for owners of small or moderate telescopes. But then another surprise, Iapetus, a thousand miles across and 8 million miles from Saturn; it goes round the planet in 79 days and its brightness is variable. When west of Saturn it is easily seen but when east of Saturn it drops below 12th magnitude. Like almost all major satellites, Iapetus has captured rotation, like our Moon, so the orbital and rotation periods are equal, Iapetus keeping the same face toward Saturn all the time.

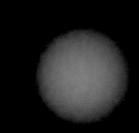

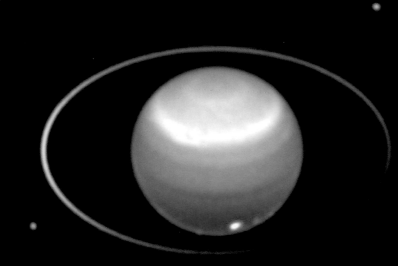

Part of the surface is bright as ice and another part is almost pure black. This baffled astronomers, but now we realize that the clue to the black and white surface lies much further out, with another satellite, Phoebe, discovered by W.H. Pickering in 1898. It is responsible for a very tenuous ring quite beyond the range of telescopes of amateur size, and was detected initially by its infrared radiation. It appears that Phoebe itself moves round Saturn in a retrograde or wrong-way direction like a car going the wrong way round a roundabout. What happens is that particles are knocked off Phoebe by meteoroids, and this dust spirals inwards, some of it landing upon the leading hemisphere of Iapetus and darkening it. This is why Iapetus has its unique yin-yang aspect.

But the mysteries do not end here. Running along its equator Iapetus has a huge mountain ridge of unexplained origin. Iapetus is indeed an interesting satellite, and we await new data from future probes.

Imaging Saturn's Moons

High-frame-rate cameras on small instruments can also be used to image the planet and its moons, again using the techniques outlined for Jupiter. Interactions between Saturn's brighter moons and the planet are less common than in the Jovian system because Saturn's rotational axis is tilted by 26.7°. From Earth we only get to see these events when Saturn is close to an equinox. This occurs approximately 14–15 years apart and at such times the planet's wonderful ring system appears edge-on to us.

Most of the larger moons are quite tricky to image as they cross Saturn's disc, and their shadows, although quite dark, are much less distinctive than those cast by the Galilean satellites on to the cloud tops of Jupiter. One notable exception is when Saturn's largest moon Titan casts its shadow on Saturn. The large dark spot that is formed when this happens is quite noticeable.

With the naked eye, Saturn appears as a brightish, slightly yellow star. Only when a telescope is used can its full beauty

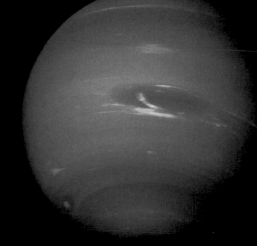

be appreciated. The planets Jupiter, Uranus and Neptune also have rings, but they are dark and difficult to detect. The bright icy rings of Saturn are in a class of their own.

All the planets we have discussed are easy naked-eye objects, and some, particularly Venus and Jupiter, are strikingly brilliant. Beyond Saturn there are two more giants but they are much fainter and further away.

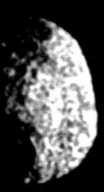

29

Uranus

First comes Uranus, discovered by William Herschel in 1781. Herschel, by profession an organist, was not looking for a planet and did not recognize it for what it was. In his paper to the Royal Society he gave an "account of a comet". Only when its orbit was calculated did it become clear that the object was a true planet and a giant one. It takes 84 years to go round the Sun and is just visible without optical aid provided you know where to look.

Its diameter is approximately 30,000 miles, less than half that of Saturn, and ordinary telescopes show little other than its pale green disc. It is a giant but not a gas giant similar to Jupiter and Saturn. Most of its material is in the form of ices, in a hydrogen-rich atmosphere.

Whether there is a solid nucleus is uncertain. Jupiter and Saturn do have solid cores – both emit far more radiation than they could if they just received radiation from the Sun alone – and the same is true of Neptune. But Uranus appears to have very little by way of internal heat, which means its make-up is probably different.

Uranus is peculiar in another way. Most planets spin with their rotation axes at a reasonable angle to the orbital plane. In the case of the Earth, our axis is tilted by 23.5 degrees, while with Jupiter it is only a few degrees and the planet moves in an upright pose. With Uranus the tilt is 98 degrees, more than a right angle so the rotation is technically retrograde or backwards. The rotation period is just over 16 hours. The orbital period is 84 years and we see Uranus through all kinds of angles. Sometimes we are facing a pole while at others the equator is presented. The reason for the tilt is unclear. It used to be thought it was hit by a massive body and tipped on its side, but this does not seem very plausible bearing in mind we are dealing with a giant of considerable size and mass. Today it is generally believed that interactions with other planets are responsible, but we still do not know why it is inclined at such an angle.

Like the other giants Uranus has some satellites. Only five are in the range of telescopes used by amateurs and all are less than 1000 miles in diameter. The most interesting is the closest-in, Miranda. It has a very varied surface – mountains, craters and ridges: a geologist's paradise.

Neptune

When Uranus was discovered, immediate efforts were made to work out its orbit and the way in which it might be expected

to move. Orbital period was easily fixed as being 84 years but as time went by, it became painfully obvious that something was wrong, and that Uranus was not behaving exactly as the mathematicians had forecast. Slowly but surely it wandered away from its calculated path until the errors became too great to be ignored. What caused them?

One suggestion was that Uranus was being pulled out of position by another large planet at a still greater distance from the Sun. In the mid-19th century several astronomers began to look at the problem. One was the Cambridge graduate John Couch Adams. By following the position of Uranus he could calculate where the perturbing planet was. In 1846 he came out with what he thought was an accurate position and sent his calculation to the Astronomer Royal, Sir George Airy at the Royal Observatory, Greenwich. Airy was not impressed and no immediate search was made.

Meanwhile there were developments in France. A Paris mathematician, Urbain Leverrier, also became interested in the problem and began working on the same lines as Adams. Note there was no contact between the two and they had never heard of each other. By 1846 Leverrier too had his proposed position and contacted the Paris observatory. Again, nothing was immediately done. Patience, however, was never Leverrier's strong point – one of his compatriots said he was not the most detestable Frenchman, he was the most detested.

Before long Leverrier became tired of waiting and contacted Johann Galle, who worked at the Berlin observatory. Galle was enthusiastic and it so happened there was a telescope suitable for planet hunting. Galle went to the observatory director, Encke, and requested to begin the search. Encke was suspicious, but ultimately decided to cooperate, saying: "Let us oblige the gentleman from Paris." A young astronomer, Heinrich D'Arrest, overheard the conversation and asked to be allowed to join in.

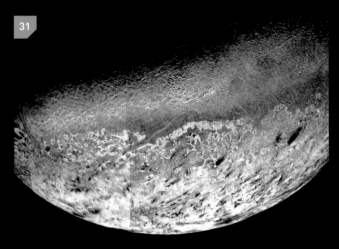

31

Galle and D'Arrest lost no time and on the first clear night went to the dome to check stars near the position Leverrier had given. Galle used the telescope while D'Arrest checked the map. A new map had recently been prepared and had not been widely distributed. D'Arrest discovered a star not on the map in only minutes. Both astronomers looked carefully and fancied they could detect a small disc – which no star can show – and they followed the body until it set. The next night it had moved by exactly the amount to show it was a planet. A triumph of ingenuity.

Remember that the hunt at Berlin was undertaken purely on the basis of Leverrier's work: the observer had never heard of Adams. By this time Sir George Airy had been roused to activity and instructed one of his astronomers, James Challis, to look for the planet. Challis had no up-to-date maps, and adopted a laborious method of checking, but did indeed find an object with a small disc. He did not call in other astronomers to confirm his findings. When he did return the sky was overcast, and Galle and D'Arrest's announcement was already made.

When the discovery of the new planet, Neptune, was announced by Sir John Herschel, the French accused the British of trying to steal the glory of the discovery. This could have led to an unpleasant incident but fortunately it did not. When Adams and Leverrier met they struck up a friendship which lasted the rest of their lives, even though neither spoke the other's language.

In size and mass Neptune and Uranus are near twins. Uranus is slightly larger and slightly less massive. Neptune's

magnitude, between 7 and 8, means it is too dim to be seen with the naked eye, but binoculars show it easily, and a telescope shows a small disc that is not green like Uranus, but bluish.

In other ways, too, Uranus and Neptune differ. Neptune does not share the remarkable axial tilt of Uranus so the seasons there are more conventional. Neptune also has a source of internal heat so there is definitely a rocky core at a fairly high temperature. It too qualifies as an ice giant though it does have a substantial hydrogen atmosphere with a considerable percentage of helium.

One satellite, Triton, discovered soon after Neptune itself, in 1846, is unique in being the only major satellite to have retrograde motion – i.e. to move round its primary in the direction opposite to that in which the primary rotates. There seems almost no doubt that Triton was originally an independent body ensnared by gravity and unable to break free.

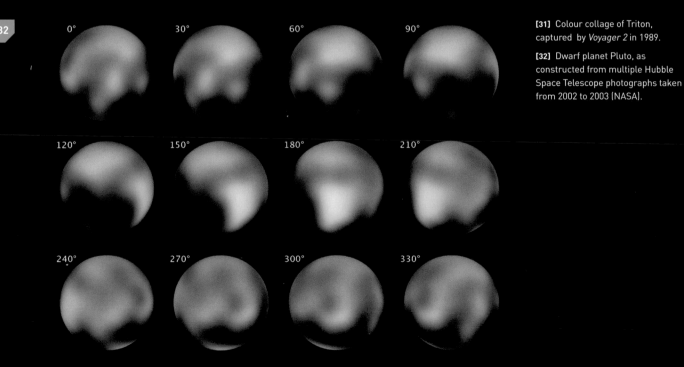

[31] Colour collage of Triton, captured by *Voyager 2* in 1989.

[32] Dwarf planet Pluto, as constructed from multiple Hubble Space Telescope photographs taken from 2002 to 2003 (NASA).

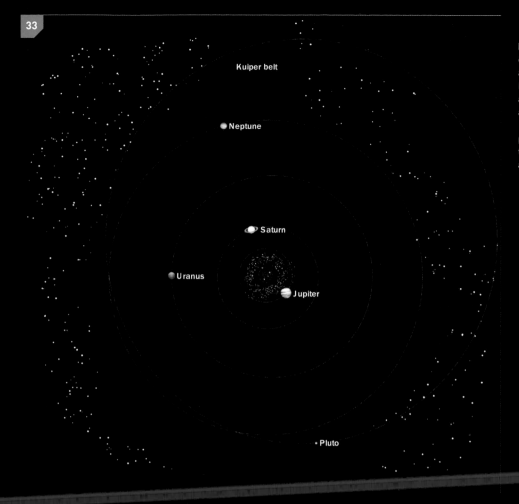

Kuiper belt

Neptune

Saturn

Uranus

Jupiter

Pluto

[33] The Kuiper Belt lies beyond the orbit of Neptune, and comprises asteroids and comets, and larger Kuiper Belt objects such as Pluto and Eris, which have been classified as dwarf planets.

The diagram also shows the position of the asteroid belt inside the orbit of Jupiter (between Mars and Jupiter).

The Kuiper Belt

We are not yet at the end of the planetary system. Out beyond Neptune we come to another swarm of asteroidal-type worlds, known as the Kuiper Belt in honour of the Dutch astronomer Gerard Kuiper, one of the first to suggest its existence.

A large member of the belt is Pluto, discovered by Clyde Tombaugh at the Lowell observatory in 1930. When found, it was assumed to be comparable in size to the Earth or at least Mars, and to be the only body of that size moving in the far reaches of the Solar System. It had a strange orbit, much more eccentric that other planets, and when at perihelion (closest to the Sun), it came within the path of Neptune. Its orbital tilt is 17 degrees so there is no fear of collision.

Also, Pluto was certainly not a giant. As better measurements were made, the diameter went down and down, and someone suggested it would soon vanish altogether. We now know its true diameter is less than that of our Moon, and Pluto cannot rank as a primary planet. In the 1990s other bodies were found – Eris is larger and more massive. Pluto is merely a Kuiper Belt Object (KBO) and only appeared bright because it is relatively near. Pluto has actually been mapped using the world's largest telescopes. It appears to have bright and dark patches on its surface but

that is all we know. A probe, New Horizons, was launched in 2006 and is currently making several close-up flybys. We will find out a great deal more about Pluto soon.

There have been major discussions about the status of Pluto: some astronomers, particularly Americans, have been reluctant to demote it from the dignity of a proper planet. However, facts cannot be denied and finally a compromise was reached. Several of the Kuiper Belt Objects have been reclassified as dwarf planets.

All are bitterly cold and we cannot expect any to retain an appreciable atmosphere. The theory that they were due to the break-up of a larger body is beset by mathematical problems. It is more likely that the objects were simply left over when the main Solar System was formed from the solar nebula.

There are some Kuiper Belt Objects that have long, narrow paths and make orbital journeys of many centuries. They might even reach the Oort Cloud, a hypothetical cloud of cometary bodies so far out that we cannot see it. Its existence is inferred.

As far as we are concerned the Kuiper Belt marks the outpost of the planetary system. There is always the chance of another major planet beyond the Kuiper Belt, but this seems very improbable.

EVENT DRIVEN
ASTRONOMY

The slender solar "horns" seen just before and after an
annular eclipse of the Sun.

CHAPTER 7

Chance alignments of objects in the Solar System can introduce a dramatic element into astronomy, with tremendous potential to create exciting images. Here we look at conjunctions, occultations, transits and eclipses, events which involve one or more bodies appearing to interact with or obscure one another. Such events only occur at certain times, something that adds another level of excitement to the proceedings.

An apparent close approach of one body to another can be quite an impressive sight. One of the most dramatic is that which occurs when the planet Venus and the crescent Moon appear close together in the sky.

Although just being close together does not fulfil the strictest definition of the term, it's common to describe the apparent pairing of one or more bodies in the sky as a conjunction. The formal definition states that they must have the equivalent of the same celestial longitude for it to be a true conjunction, but close together seems an acceptable definition for most situations.

As Venus can never venture too far from the Sun, any conjunction that happens between it and the Moon must occur when the Moon is in its crescent phase; that's an old crescent in the morning sky or a young crescent in the evening sky. This means that the event normally takes place during twilight with a blue gradient sky background adding an extra dimension to the already stunning view.

Conjunctions between the Moon and the other planets can also occur, and those between Jupiter, Mercury and, to a lesser extent, Saturn, can also be impressive sights to behold. There's no real scientific worth to a conjunction, this being an astronomical event to just savour and enjoy.

Conjunctions between the Moon and planets are quite common because all of these bodies travel across the sky close to the ecliptic – the apparent path of the Sun against the background stars. This is a great circle around the sky which has the Earth at its centre. The ecliptic can also be thought of as defining the plane of the Solar System, and the orbits of the major planets and the Moon tend to lie close to this plane. Consequently, when you see a major planet or the Moon in the sky, it will be lying quite close to the ecliptic.

From this description it's possible to work out that the major planets can also be in conjunction with one another. However, having a slower apparent daily motion against the background stars, conjunctions between planets are far less common than those between the Moon and the planets. The Moon's a fast mover and will pass all of the planets at least once every month. Catch two bright planets, such as Jupiter and Venus, close together in a relatively dark sky, and the sight can be very impressive indeed. When three or more planets or two planets and the Moon appear close to one another in the sky, this is known by the infrequently used term "a massing".

The Moon has a tangible size in the sky: an apparent diameter of around 0.5 degrees. During its monthly travels against the background constellations, the Moon's disc may sometimes pass between us and a planet, hiding the planet from view. Such an event is known as an occultation, or more specifically, a lunar occultation. Despite the Moon's apparent size in the sky, lunar occultations of planets aren't that common.

Lunar occultations of a planet are fascinating to watch through a telescope. As the Moon catches up with the planet, the edge of the Moon can be seen to cross the planet's comparatively tiny disc before covering it completely. This is what's known as an occultation disappearance event. After a period of time that depends on where the planet happens to

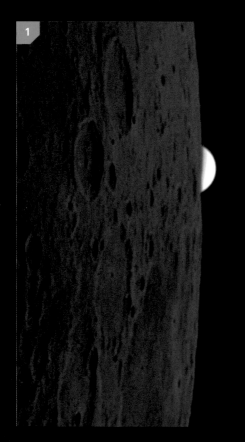

[1] Venus peeks out from behind the Moon at the end of a lunar occultation of the planet

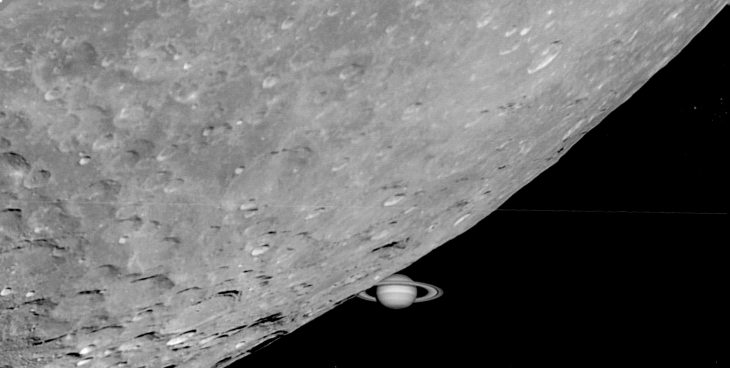

[2] A "conjunction" between the Moon and Jupiter.

[3] A crescent Moon sits near the Pleiades open cluster. An extended exposure has over-exposed the crescent but allowed the camera to record the cluster stars and the dim, earthshine lit portion of the Moon's surface.

[4] A grazing lunar occultation of the planet Saturn imaged in March 2007 from the UK.

be passing behind the Moon's disc, the planet will reappear from behind the trailing edge of the Moon. This is known as an occultation reappearance event.

A special kind of occultation occurs when you see the edge of the Moon partially clip the planet's disc but never completely cover it; an event known as a grazing lunar occultation. As an aside, this sort of event can also happen between planets, although the small size of a planet's disc compared with the relatively slow speed that planets move relative to one another means that such events are quite rare. The next such event will occur on 22 November, 2065, when Venus will appear to pass across the southern edge of Jupiter.

The Moon and planets can also occult the background stars. Again, such events are more commonly seen with the Moon than the planets although occultations of bright stars may not be as common as you would think for a body with such a large apparent diameter.

Lunar occultations of stars are fascinating to watch because the star is effectively a point source of light and the Moon's edge, bereft of atmosphere, is sharp. Consequently, the star's light extinguishes instantly during occultation disappearance and reappears instantly at occultation reappearance.

As the Moon shows phases, occultations that happen when the Moon's between new and full occur with the star or planet disappearing behind a dark lunar limb and reappearing from a bright one. Between a full and new Moon, the reverse is true.

Occultations of stars by a major planet do also happen but, again the planet's relatively small disc size means such events aren't that common and occultations of bright stars by planets are very rare. The magnitude +3.3 star Theta Ophiuchi will be occulted by Mercury on 4 December, 2015. Magnitude +2.9 Pi Sagittarii gets the same treatment by Venus on 17 February, 2035, while brilliant Regulus in Leo is also occulted by Venus on 1 October 2044.

A more common planetary occultation occurs when a major planet with moons appears to hide one of its moons. This can be seen to occur with Mars, Jupiter, Saturn, Uranus and Neptune, although only Jupiter shows events that are fairly common and easily seen with a small telescope. Occultation of Saturn's brighter moons by the planet can only occur when Saturn's tilt angle is fairly edge-on to Earth, something that occurs at approximately 15 year intervals. The two moons of Mars are too faint for such events to be seen convincingly with amateur equipment, the same being true for the brighter moons of Uranus and Neptune.

If the body that's passing in front of another is too small to occult the more distant one, the event is called a transit. Transits are quite common in the Jovian system where the four Galilean moons may be seen to pass in front of Jupiter's disc. Such events are common for Io, Europa and Ganymede but less so for the slower moving and more distant Callisto.

Transits of the main Saturnian moons may also be seen as long as the planet's tilt is such that the Earth is more or less in its equatorial plane. Such events are much harder to see

[5] The brilliant planet Venus in conjunction with the Pleiades open cluster.

[6] Planet Mars close to the open cluster Messier 35 in Gemini.

[7] Venus transiting the Sun's disc in June, 2004.

[8] An annular eclipse of the Sun photographed at the point of maximum "annularity". The Moon's disc is perfectly concentric within the larger disc of the Sun. This is known as the "ring of fire".

than those which occur in the Jovian system, requiring a large telescope to view convincingly.

Having orbits which are smaller than the Earth's, the inner planets Mercury and Venus may be seen to transit the Sun. Transits of Mercury occur 13 or 14 times per century while those of Venus are much rarer, occurring in pairs separated by 8 years but with a long wait of 105.5 or 121.5 years between pairs. The whole 243-year cycle then repeats. The next transit of Venus occurs in 2117 with another in 2125.

Transits of Mercury can only occur in May and November. The nearest transits of Mercury to the time of writing this

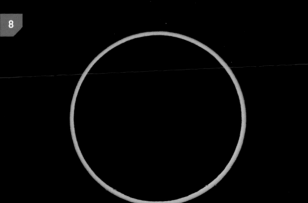

book will occur on 9 May, 2016, 11 November, 2019, and 13 November, 2032.

Sun-Moon Interactions

The Sun and the Moon have similar apparent diameters in the sky, but the elliptical orbit of the Moon around the Earth coupled with the elliptical orbit of the Earth around the Sun means that the apparent size of both bodies does vary slightly. If the apparent size of the Moon is smaller than normal and the size of the Sun is larger, it's possible for the Moon to appear to pass in front of the Sun's disc without totally covering it. If you're standing in the right place on Earth, the outer edge of the Moon appears to be totally surrounded by the edge of the Sun. This transit event is also known as an annular eclipse of the Sun. The appearance of the Moon's dark silhouette central within the Sun's disc gives rise to what's known as a "ring of fire".

A more spectacular form of solar eclipse occurs when the apparent size of the Moon's disc is greater than that of the Sun. If this occurs and you happen to be standing in the right place at the right time, you'll see the Moon's disc completely cover the Sun's disc in an event known as a total eclipse of the Sun.

The term "eclipse" is used to describe an event where an

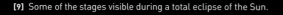

[9] Some of the stages visible during a total eclipse of the Sun.

object is made to disappear from view. From this definition, an occultation can also be described as an eclipse, as can the event that occurs when one body is hidden by the shadow of another.

As stated earlier, the apparent diameters of the Sun and Moon in the sky are similar but subject to slight variation due to changes in distance between the Sun and the Earth and the Earth and the Moon. These changes occurring because the orbits concerned are elliptical rather than circular. If both the Sun and the Moon line up in the sky there are three possible size combinations:

- The Moon appears larger than the Sun
- The Moon appears smaller than the Sun
- The Moon appears exactly the same size as the Sun

Every body in the Solar System that is bathed in the Sun's light casts a shadow in the opposite direction. As the Sun is not a point source at the distance of the major planets, the shadows cast take on a very specific three-dimensional form. A dark, converging shadow cone extends away from the body

[10] Dappled sunlight through trees takes on the shape of the crescent Sun during the partial phase of an annular eclipse.

surrounded by a diverging less intense region of shadow.

If you are in the dark central shadow cone and look back towards the Sun, it would be completely hidden, none of its light managing to get to your eyes. Outside of the dark cone in the diverging, less intense shadow, part of the Sun's disc would be visible. The inner dark cone is known as the umbral shadow while the outer, less intense region is known as the penumbral shadow. From inside the umbral shadow, the Sun would appear totally eclipsed while from within the penumbral shadow it would appear partially eclipsed.

The intensity of the penumbral shadow increases the closer you get to the inner umbral shadow. This is because more and more of the Sun's disc appears partially eclipsed. Outside the penumbral shadow, the Sun's disc would be totally clear from obstruction and the Sun would appear with full intensity.

The Moon is approximately 1/400th the size of the Sun and, by a remarkable coincidence, the Sun is approximately 400x further away than the Moon. This is the reason why both bodies have very similar apparent sizes in the sky. The Moon's umbral shadow extends out to a distance similar to the distance between the Moon and the Earth. For the situation where the Moon and Sun have the same apparent diameter in the sky, the umbral shadow cone tip just reaches the Earth's surface. If you were to stand in the correct location to experience the shadow tip pass over you, the Sun would appear to undergo an almost instantaneous total eclipse under these conditions.

For the situation where the apparent Moon's disc looks smaller than the Sun, the dark shadow cone tip isn't able to quite reach to the Earth's surface. If you positioned yourself accurately, using suitable protective filters, at the time of maximum "eclipse" you'd see the Moon's disc concentrically positioned within the Sun's disc – the event described earlier as an annular eclipse. In the strictest sense this isn't an eclipse at all, it's a transit of the Sun by the Moon, but the term eclipse is still used.

Finally, if the Moon's apparent disc happens to appear larger than the Sun in the sky, if you're in the right place on the Earth's surface at the right time, you will witness that most sought after of astronomical events, a total eclipse of the Sun. Here, the tip of the dark shadow cone would lie below the surface of the Earth. As the tip of the cone is truncated, on the Earth's surface the shadow has a tangible size and if you're located so that it passes over you, you get to see a total eclipse lasting minutes.

The size and shape of the shadow, as well as how fast it appears to move across the Earth's surface, is determined by the relative apparent sizes of the Sun and the Moon along with the geometry of the eclipse relative to the Earth. Shadows cast to the edge of the planet give rise to more distorted and faster moving ellipses.

The largest umbral shadow has a width of around 150 miles and the speed at which the shadow races across the Earth's surface varies from up to 1,100 mph at the equator to 5,000 mph at the poles. The shadow moves across the surface in a west to east direction. The longest duration that a total eclipse of the Sun can last is around 7.5 minutes. You need to be standing in the path of the track in order to experience the full majesty of a total solar eclipse. Observers either side of the track experience a partial eclipse of ever decreasing magnitude the further from the main track they are located. A typical eclipse shadow is 100

miles wide and travels 10,000 miles across the surface of the Earth, covering less than one percent of the Earth's surface as it goes.

As stated, the main types of solar eclipse are partial, annular and total, but the fact that the Earth's curving surface presents a projection screen of varying distance for the shadow gives rise to another type of solar eclipse, known as a hybrid. The term refers to the situation where the dark shadow cone tip can't quite reach the Earth's surface at the track start but as our planet's surface curves upward towards it, it can. So at the start of the track, the eclipse would appear annular but further along the track it becomes total. Finally, as the Earth's surface curves away again, the eclipse can return to an annular type.

Of course, it's impossible to view all of the stages because of the speed that's required to chase the shadow, but at least, with forward planning, you can choose whether it's an annular or total solar eclipse you want to watch. For reasons that should appear obvious after reading the description of a total solar eclipse below, most people opt for the total solar eclipse.

A Total Eclipse of the Sun

Assuming you have done your homework and positioned yourself in the right place at the right time and the weather is being kind, it is possible to witness the majesty of a total eclipse of the Sun. The eclipse starts as the Earth enters the penumbral shadow of the Moon. At this point, the Sun can only be viewed with a certified protective solar filter. Using a filtered telescope, the first evidence that something will happen is a tiny "dent" at the edge of the Sun, the tell tale evidence that the hitherto invisible Moon is indeed in the predicted position. This event is called first contact.

Over the tens of minutes that follow, the Moon's disc gradually encroaches over the Sun, leading to an ever increasing partial solar eclipse. Amazingly, it's not until 90 percent or more of the Sun's disc is covered, that there's any noticeable change to the surrounding light level. Then after what seems to have been a rather lengthy wait for the partial eclipse to become almost

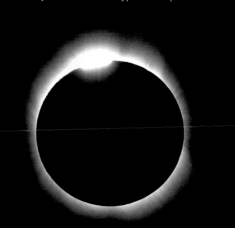

11

[11] The so called "Diamond Ring" effect caused by sunlight shining through depressions in the edge of the Moon. There are two diamond rings during a total eclipse – one just before totality begins and one that occurs just as it ends.

[12] Third contact Diamond Ring effect. The instant when the total phase of the eclipse ends.

[13] Long exposures combined with shorter exposures to reveal the corona surrounding the eclipsed Sun. Note the features visible on the Moon's earthshine lit face using this technique.

[14] The Earth's shadow covers the Moon during a total lunar eclipse. If you were on the Moon's surface looking back towards the Earth, you'd see the light from the Sun forming a red ring around the edge of our planet. This is caused by the Earth's atmosphere scattering blue light and refracting (bending) the remaining red light to effectively infill the dark shadow. From the Moon's surface you'd witness a ring of sunrises and sunsets all happening at the same time.

total, everything suddenly starts to happen at once!

As the initial partial phase progresses to totality, the remaining bit of Sun takes the form of a thin crescent. Sunlight passing through small apertures such as a pinhole in a piece of card, project the shape of this crescent onto the ground. Even dappled sunlight passing through the leaves of a tree shows this effect.

As the crescent gets ever thinner, shadows take on a strange appearance, looking sharp in one direction but distinctly fuzzy at right angles to that direction. This is because the Sun's thin crescent acts like a slit of intense light, producing sharp shadows in the direction of the slit's shortest dimension. Along the length of the slit, it's as if you have lots of point sources of

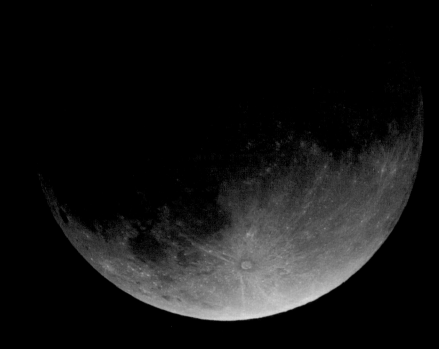

light all lined up. Consequently, the shadows cast look fuzzy in this direction.

The effect gets more pronounced as the crescent starts to get very thin and it's at this time that other physical effects make themselves known. The light from the Sun takes on a distinctly twilight feel, dimming ever quicker as totality approaches. The temperature also drops noticeably at this time, and animals may react strangely at this point too.

As the Moon makes its final move across the solar disc, the demise of the thin crescent quickens, its horns accelerating towards one another until there is just a hint of sunlight left visible at the edge of the Moon. On the ground the strange phenomenon known as "shadow bands" may be seen. Caused by the thin strip of sunlight being affected by the Earth's moving atmosphere, the bands appear like undulating thin waves of light and dark that move and shimmer in parallel. These are best seen against a plain background such as a white bed sheet.

As this time, the remaining sunlight just visible at the edge of the Moon pours through irregularities on the Moon's edge caused by mountains, crater rims and valleys seen in profile. The brilliance of the Sun looks intense as it appears through the depressed regions, producing bright spots of light known as Bailey's Beads.

The appearance of the beads is coupled with a marked darkening of the sky around the Sun's disc, a time when the inner part of the Sun's atmosphere, the chromosphere and solar corona, start to become visible. At this point in the proceedings the edge of the Moon is surrounded by a glowing ring of light and with the Bailey's Beads resembling a brilliant jewel, this is the first so called "diamond ring" of the eclipse.

The drift of the Moon continues until the Bailey's Beads are also extinguished. After this time, the eclipse must be viewed without a filter to get the best from it. With the main disc of the Sun covered, features which are normally hidden from direct vision become visible. One of the most striking is the appearance of the chromosphere, a layer of hot hydrogen gas that sits above the normally visible surface. This appears with a striking reddish-pink colour around the Moon's edge. The term "chromosphere" literally means "sphere of colour".

This is a time when it may also be possible to see prominences reaching off the edge of the Sun too. These have the same red-pink colour as the chromosphere and a large prominence around the edge of the Sun is quite a sight to behold.

As your eyes slowly get used to the reduced light levels, the dim glow of the corona begins to extend visually away from the Sun's disc. The appearance and form of the corona is magnetically influenced and varies considerably depending on the current activity level of the Sun. The longer totality lasts, the

[15] Comet C/2006 P1 McNaught photographed against a bright twilight sky in January 2007.

[16] Donati's comet over St Paul's cathedral in London, 1858. From Agnes Giberne's book Sun, Moon and Stars published in 1884.

[17] Depiction of a cometary nucleus warming as it gets closer to the Sun. Material is vented off the surface, and as the nucleus rotates, this material forms a glowing shroud of gas and dust that forms the comet's head or coma. Radiation pressure exerts force on the dust particles pushing them away from the Sun where they spread out to form a curving dust tail. Electrically charged particles are carried back by the solar wind to form a long straight gas or ion tail.

more accustomed to the view your eyes become and eventually, you may see the corona extended away from the Sun for many solar radii.

The temperature at this time drops noticeably and standing in the shadow of the Moon looking up, it's a rather humbling experience realising just how reliant we all are on the output of the Sun.

Finally, after what seems like a ridiculously short period of time, the bumpy edge of the Moon once again starts to let sunlight through. Bailey's beads once more appear along with a second diamond ring. This one is often more dramatic than the first as your eyes will have become partially dark adapted during totality. After a brief view, it's time to don the protective filters once again as the end of the eclipse plays out as a reverse of the first half. If your location is such that you have a good flat view around you, it's possible to sense the approach and despatch of the Moon's shadow before and after totality.

We are very fortunate to be able to witness this amazing event. The Moon's distance from the Earth is slowly increasing over time. Hundreds of millions of years in the past, the Moon would

have appeared too big for the precise fit that's required for a proper total eclipse of the Sun. Similarly, it's estimated that in 600 million years time, the Moon will appear too small for a total eclipse to be seen at all. During the course of a year, there will be between two and five solar eclipses, though of course seen from different locations on Earth. Not more than two will ever be total eclipses, however.

Lunar Eclipses

The counterpart to a solar eclipse is the lunar eclipse, an event which occurs when the Moon lines up with the Sun on the opposite side of the Earth. When this occurs, the Moon will pass through the shadow of the Earth and what was a bright full Moon will appear dim by comparison.

The Earth casts a shadow in space which is similar in make-up to the one described for the Moon above. There's a dark umbral shadow and a lighter surrounding penumbra. If you are in the umbral shadow of the Earth looking back at the Sun, the Sun's light would be totally blocked. Inside the Earth's penumbral shadow looking back at the Sun, part of the Sun's disc would be blocked.

The Earth's umbral shadow is larger than that of the Moon and at the distance of the Moon is still quite large, presenting a larger target for the Moon to pass through. Unlike a solar eclipse, a lunar eclipse can be seen from anywhere on Earth as long as the Moon is above the horizon.

The two main types of lunar eclipse are partial and total. A third type is known as a penumbral eclipse but this is a tricky thing to see visually unless the Moon's passage through the penumbral shadow takes it close to the umbral shadow. When this occurs, one edge of the Moon may appear to be slightly more shaded than normal.

You might expect the Moon to go totally dark during a total lunar eclipse but this is not the case. The reason for this is the Earth's atmosphere. As light passes through the atmosphere it gets bent, or refracted, so that it in-fills the umbral shadow. As the bluer components of this light get scattered by the atmosphere, the colours left are towards the red end of the spectrum. The Earth's umbral shadow is quite reddish in colour as a consequence of this.

As the Moon enters the umbral shadow it appears to change colour until at the time of totality it will look quite reddish. The darkness and depth of the colour depend on the state of the Earth's atmosphere. Volcanic eruptions and heavy cloud cover can all have an effect. Some lunar eclipses take on a light gold-coppery hue, while others can go a deep blood red, the Moon virtually disappearing from view.

The pace of a lunar eclipse is in complete contrast to the frenetic activity surrounding totality of a total solar eclipse, giving you ample time to go out and enjoy the event.

Comets

A comet is quite unlike a planet. It consists of a small icy nucleus, surrounded by dust and gas. Some comets move round the Sun in short periods, so we know when and where to expect them. Others come from the depths of space and take us by surprise, paying us only one visit before retreating to the distant parts of the Solar System. Much the most famous of all is Halley's Comet, named after the well known English astronomer Edmond Halley, which he did not discover but whose orbit he was the first to calculate. It has a period of 76 years and can become a brilliant naked eye object. At its last return in 1986 it was badly placed and not very conspicuous. We know a lot about it because unmanned spacecraft visited it. One penetrated the heart of the comet until it was put out of action by small particle – probably no larger than a grain of rice. A comet has no light of its own and depends upon what it receives from the Sun.

Short period comets go round the Sun in periods of a few years – in the case of Encke's Comet, 3.3 years – but not many of these are visible with the naked eye. Every time a comet passes its perihelion or closest approach to the Sun it loses a certain amount of material and this wastage cannot go on indefinitely so by astronomical standards comets are short lived. Several comets that used to return regularly to the Sun have now vanished completely. Large comets have tails, which may be of two types, a curved tail made up of dust, and a straight tail made up of ionised gas, all robbed from the nucleus.

We never know when a really bright comet may appear, and when they do so they cause a great deal of interest, and

in former times, fear, as comets were regarded as portents of doom. "When beggars die there are no comets seen, the heavens themselves blaze forth the death of princes"– Shakespeare.

Over the centuries there have been many alarms raised about possible comet collisions, and it is quite true that the impact of a nucleus just a few miles across could have devastating effects, but in the recent past, this has not happened.

As a comet warms in the Sun's light, the outgassing that follows causes a trail of debris to be left along its orbit. If the Earth's orbit encounters this debris, the typically sand grain-sized particles crash into Earth's atmosphere where they vaporize, leaving a streak of light known as a meteor.

Meteors and Meteorites

The term shooting star refers to the brief streak of light that you sometimes see as a small particle known as a meteoroid vaporizes in our atmosphere. Known formally as meteors, they are nothing to do with real stars of course, and occur much closer to home, typically at heights between 45-70 miles up. Randomly occurring meteors are known as sporadics while those which appear because the Earth moves through the dust scattered around the orbit of a comet, can appear more organised and produce what's known as a meteor shower.

Annual meteor showers of note are the Perseids, which peak around the 13th of August, and the Geminids, which peak around the 13th of December. A typical meteor shower lasts for several days if not weeks but peak activity only tends to happen

[18] A Perseid meteor imaged sweeping across the constellation of Cygnus.

[19] A Geminid meteor.

[20] Shower radiant: the meteors in this shower appear to originate from one place in the sky (Chen Huang-Ming).

[21] Metorite found in 1516.

[22] Part of the Barwell Meteorite.

around a narrow window often just hours in width. A perspective effect makes it seem that shower meteors come from the same general point in the sky, known as the shower radiant. Over the period the shower is active, it's not unusual for the radiant position to drift slightly in the sky. The constellation or star that the radiant is closest to when the shower's activity is at its peak tends to be the driving force behind naming the shower. So, for example, the peak output of the Perseids occurs when the radiant is in the constellation of Perseus.

The number of meteors a shower can produce is known as its zenithal hourly rate or ZHR and this is a figure, which can lead to confusion. The ZHR of a shower is the number of meteors you'd see under perfect conditions with the radiant directly overhead at the zenith and assumes you could see the whole sky in one go. These conditions are rarely met, and the visual rate is often significantly lower than the quoted peak ZHR which should be used as a relative indicator of how active the shower is, rather than what you're likely to see.

The peak ZHR for the Perseids is between 80-100 meteors per hour, while that of the Geminids is 120. Some showers exhibit variable peak ZHRs. One notable example is the Leonid shower which shows storm level displays roughly at 33-year intervals. The last spectacular displays were at the start of the current century when thousands of meteors per hour were seen around the night of 17/18 November.

The streak a meteor makes as it passes across the sky is known as a meteor trail. After a bright meteor event, it's sometimes possible to see a ghostly glow along the trail path, a bit like a short section of a plane's vapour trail. This is caused by a decaying column of ionized gas and is known as a meteor train. Trains that last for several seconds may be subject to the effects of high altitude winds and be seen to twist and distort.

Meteors vaporise away at least 40 miles above the ground and so can do us no damage, but now and then Earth is struck by a larger body which may weigh tonnes. This is known as a meteorite and is not connected with meteors. They come from the asteroid belt and there may be no essential difference between a very large meteorite and a small asteroid.

Meteorites are generally stone, iron or a mixture. Hunting for them is an engaging hobby, but beware, a meteorite does not always betray its nature at first sight. In the 1980s a meteorite fell in Barwell in Leicestershire. Patrick Moore knew where it had come in and calculated where to find it. He went to see the local farmer and requested permission to look in his fields. He was amenable and before long Patrick found an excellent meteorite fragment, which now sits in his study, and is bequeathed to a museum.

There are other problems too. Patrick once produced a television show about meteorites and mentioned that some people like to search for them. People sent in their finds, and one particularly intriguing meteorite turned out to be a fossilised Bath bun. The largest known meteorite is still lying where it fell at Hobart West in Africa – weighing over 60 tonnes.

[23] Meteor Crater in Arizona averages 1.2 kilometres in diameter. Its scale can be appreciated by looking at the size of the visitor centre on the right hand part of the rim (Meteor Crater Arizona).

[24] Wolfe Creek Crater in western Australia is on average 875 metres in diameter (Patrick Moore).

Others have fallen and produced craters, essentially similar to those on the Moon, but smaller. The finest is Meteor Crater in Arizona. Fifty thousand years ago the meteorite struck Earth and left the giant crater. It is easily reached from Highway 99 and there is a museum on site. It was once possible to scramble down the wall into the crater, but this was stopped when one visitor fell and injured himself. Patrick did one TV show there and was dropped into the crater by helicopter, entirely alone in the mile-wide crater – the atmosphere was eerie. There is no crater quite as impressive as this one. At one stage efforts were made to dig up the meteorite but most of it was probably vaporized away during its death plunge.

Another meteor crater, Wolfe Creek in the Northern Territories of Australia is in a very deserted part of the country. However, once you get there it is well worth seeing.

We have to admit that the impact of a large meteorite say half a mile across would cause immense damage if it landed in a populated area and could wipe out a city.

There must have been major impacts in the past and there is wide support that 65 million years ago the meteorite that landed in the Yucatan peninsula changed the climate to such a degree that the dinosaurs died out.

Meteors are relatively easy things to photograph with a DSLR camera as long as luck is on your side. A DSLR set on a standard tripod using a relatively wide-angle lens and a shutter release cable is all you need. The lens should ideally be 50mm or shorter in focal length. Typically, something around the 18-28mm focal length is ideal for a non full-frame DSLR camera. The ISO needs to be set high, typically 1600-3200 and the lens aperture fully opened. The faster the lens, the better and f-ratios of f/1.8 – f/2.8 are recommended if available. Prefocus the lens at infinity and turn auto-focus off. Set the exposure time to say 30s and set the camera's shooting mode so that it repeats shots if the shutter button is held down

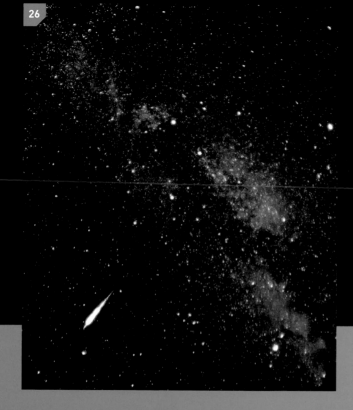

(burst shooting). Lock the remote shutter release in the on position and sit back and wait.

The best place to point the camera is typically about 60° up in the sky and roughly 45-60° to either side of the radiant. Actually, this is less critical and great shots can be had by pointing the camera at the radiant itself. However, trail lengths appear longer the further they occur from the radiant. Consequently, any shower meteors that are caught close to the radiant will appear quite short.

You'll need a decent sized memory card to store all the photos you take throughout a meteor photography session and make sure you have plenty of charged camera batteries on standby too. Finally, when done, the only way to check whether you've been successful is to download the shots to a computer and sift through them one by one. Laborious though this is, just one bright meteor in a shot can make the whole task completely worthwhile!

[25] A bright Perseid Meteor streaks through the sky near the Pleiades open cluster.

[26] A Kappa Cygnid meteor streaks across the sky.

CITY OF
THE STARS

CHAPTER 8

By "city of the stars" we mean our own particular Galaxy, the Milky Way, though technically the Milky Way is the main plane of our Galaxy where a great many stars lie in roughly the same direction. When we look up into the night sky with the naked eye we see the 88 star patterns, or constellations, that fill the sky. The constellations were originally maps made by astronomers in ancient times.

Once we go beyond the Solar System, our own particular part of the Universe, we have to deal with distance of quite a different order. This is where we use light years rather than miles or kilometres, which are far too short. An important point here is that no one can really understand these huge distances. Can you picture a million miles, let alone 24 million million – the distance to the next nearest star to the Sun? We certainly cannot and it's virtually impossible to put these distances meaningfully into plain language.

The distance to the nearest star, Proxima Centauri, becomes somewhat trivial as you head out into the deepest depths of space. At 4.2 light years distant Proxima is a near neighbour in the gravitationally bound island city of stars that we call our Galaxy. Measuring approximately 100,000 light years in diameter, the best model for the Milky Way is to think of two fried eggs slapped back to back. The bulge in the middle represents the core of the galaxy, while the flatter whites are where the spiral arms are located – vast regions of star birth. The stars in the arms are younger than those at the core so from a distance, the core of a spiral galaxy such as ours appears redder than the hot blue stars which give the

spiral arms their form.

Our Sun lies approximately two-thirds of the way from the core towards the edge of the galaxy. All of the individual stars we see in the night sky lie within our own galaxy in a "bubble" of visibility that is approximately 10,000 light years in diameter. There are many other stars – several hundred billion in total in the Milky Way Galaxy alone – lying outside of this bubble that are too far to be seen individually with the naked eye. The light from these distant suns merges together to form the misty path that we sometimes see running across a dark moonless night.

It is amazing to reflect that some people who live in city centres have never seen the Milky Way at all. Not long ago a party of school children from inner London came to Patrick Moore's observatory and when shown the Milky Way crossing the sky they were incredulous. Light pollution is a serious problem and although some councils are doing their best to limit the damage it is fair to say there are not many areas left into which artificial lights do not intrude in some way or other.

There are many different types of stars in our galaxy, including so-called main sequence stars like our Sun, huge

1

[1] The star Sirius (the "Dog Star") is the brightest night time star in the entire sky. It lies in the constellation of Canis Major, the Great Dog.

[2] Planetary nebulae, seen by the Hubble Space Telescope. NASA.

[3] Sirius is the brightest star in the sky in this photograph taken from the Canary Islands by Nik Szymanek.

[4] The Dumbbell Nebula (Messier 27) is located in the constellation of Vulpecula, the Fox. (RGB filtered mono-CCD camera image.)

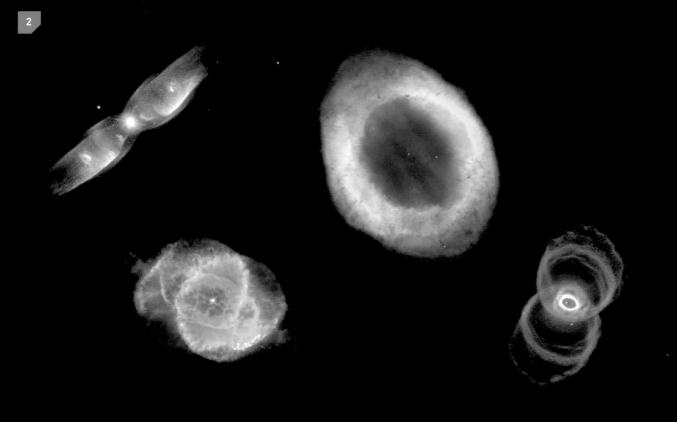

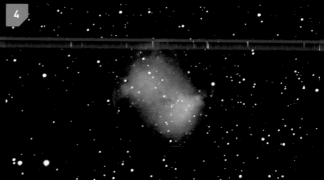

red giants, supergiants and hypergiants. At the other end of the scale there are stars that are much smaller than our own Sun, such as white dwarfs, red dwarfs and exotic objects known as neutron stars. There is a tremendous range in luminosity too and with the naked eye one can see stars that are many thousands of times brighter than our own Sun that appear like dim twinkling points simply because they are very distant indeed.

The Dog Star

The brightest star of them all in Earth's night sky is Sirius which lies in the constellation of the Great Dog (Canis Major). Intense though it looks, the brilliance of Sirius largely comes about because it is actually one of our nearest neighbours. When seen low, near the horizon, Sirius twinkles a furious range of colours but this is simply an atmospheric effect and in reality, the star is pure white.

Let us run over the life story of a star like our Sun. It begins by condensing out of the material in a gaseous nebula, a process that takes some time. As the fledgling star

becomes smaller and denser under the influence of gravity, the temperature rises until it reaches 10 million degrees. This is where nuclear reactions begin, and hydrogen atoms combine to form helium, releasing energy and losing mass. The internal energy helps inflate the star until an equilibrium state is reached. At this point, gravitational collapse is balanced against internal energy release. This is the most stable period in the life of a solar type star and can last for thousands of millions of years. But the supply of hydrogen is not inexhaustible, and there comes a time when it starts to run low.

The star then has to change its whole structure. Heavy atoms are built up from those already created, until finally, there comes a point where internal energy producing reactions are no longer possible, and the star loses the ability to keep itself inflated against gravitational collapse.

What happens then depends upon the initial mass of the star. In a modest star like the Sun, the core shrinks due to gravity and this causes layers of remaining hydrogen close to the core to react violently. The energy produced causes the star's outer layers to inflate, producing what's known as a red giant.

Eventually, these outer layers are blown away, producing what we call a planetary nebula. This is a slightly misleading term because a planetary nebula is not really a nebula and has nothing to do with a planet! The end product is a very small, very dense star known as a white dwarf, which shines feebly due to continued contraction. In the end it can contract no more and becomes a cold, dead world, still orbited by the ghosts of its remaining planets. A white dwarf is a very dense object and a teaspoon of white dwarf material on Earth would weigh about the same as a family saloon car!

Supernova

However, consider a star say 10 times as massive as our Sun. It is created inside a nebula as before, heats up and begins to shine. Everything happens at an accelerated rate. The supply of hydrogen will not last for nearly so long as with a star such as the Sun, and when the crisis comes it is of a very different order. The star collapses. There is an implosion followed by an explosion, and the star blows most of its mass away into space. This is called a supernova outburst. For a brief period the luminosity can be over a 100 million times that of the Sun but dies down fairly quickly and you are left with a very small super-dense object known as a neutron star.

A neutron star is a very strange object which has all of the space inside its atoms removed. Consequently the star is quite small, typically having a diameter about the same size as a large city. A teaspoon of neutron star material would weigh more than Mount Everest!

Neutron stars have intense magnetic fields and material that is caught by these fields radiates brightly as it is dragged down to the surface of the star. Neutron stars can rotate very rapidly indeed and if one happens to rotate in such a way that we get to see its glowing magnetic poles pass our

line of sight, the star is seen to pulse. The most famous example of this type of object, which is known as a pulsar, is that which sits at the heart of the supernova remnant known as the Crab Nebula. This particular example rotates 30 times per second and its flashes can be seen across a whole range of wavelengths including visible light. The fastest pulsar currently shows pulses at regular intervals of 1.4 milliseconds. Pulsar PSR-J0737-3039 is a very unusual object consisting of two pulsars in orbit around one another.

If the original star is large enough, the resulting neutron star may actually be too heavy to exist as such an object. If the central object has a mass over 3-4 solar masses, then its gravity will cause it to collapse completely into an object known as a black hole. This is a very exotic object that has such an intense gravitational field that light itself cannot escape from its surface.

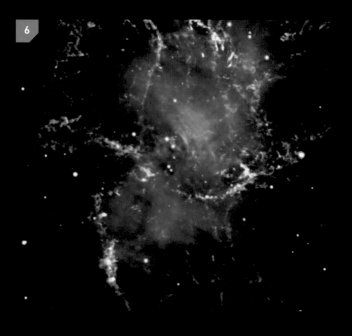

[5] Supernova 1987a is seen in the left-hand photograph, and, at right, after it had subsided. The stellar explosion that occurred here was due to a star undergoing core collapse – a Type II supernova. A Type Ia supernova occurs when a white dwarf in a binary system accretes matter from the other star, re-kindling nuclear fusion in the white dwarf. Anglo-Australian Observatory, photograph by David Malin.

[6] The Crab Nebula, seen by the Very Large Telescope (VLT). ESO.

[7] Artist's impression of the double pulsar. Michael Kramer, Jodrell Bank Observatory, University of Manchester.

Stellar Classification

The light that we receive from a star contains a vast amount of information about its make-up and its characteristics. Just by looking at a star's colour we can tell what the temperature of the star is, how big it is and how luminous it is.

Stars are categorized by a spectral classification scheme known as the Morgan-Keenan classification system or MKK system for short. The main class types are divided by letters of the alphabet in the sequence O, B, A, F, G, K and M. The stars identified by this classification scheme get progressively cooler along the scheme. Stars of class O are hot blue stars with temperatures in excess of 33,000 degrees, masses in excess of 16 solar masses and luminosities greater than 30,000 times that of our Sun. The apparently faint star Lambda Orionis, which marks the head of Orion the Hunter, is in reality nothing of the sort, being an O-type highly luminous star.

At the other end of the classification scheme, most M class stars are relatively cool red stars with temperatures less than 3,700 degrees, masses less than 0.45 solar masses and low luminosities, less than 0.08 times that of the Sun. The M classification also covers some rather large red-supergiant stars. Examples of M-class stars are the closest stellar neighbour to the Sun, the red dwarf star Proxima Centauri, and Betelgeux, the red supergiant that marks the north-eastern corner of the main pattern of Orion. One of the largest stars known, VY Canis Majoris, a red hypergiant star in the constellation of the Great Dog, also comes under the M classification.

Each class letter is subdivided further by the numerals 0 to 9, the lowest number representing the hottest member of that particular class. For example, an A0 star will be hotter than an A1 star, and so on, down to A9.

The MKK system has a further subdivision to indicate the luminosity class of the star, indicated by the Roman numerals I to V. I represents supergiants while V are dwarf or main-sequence stars. The Sun has a classification of G2V indicating that it's a G-type main sequence star.

Modern research has revealed further types of star that merit spectral classes of their own. To cater for this need, extra identifiers have been created and include amongst others; L, T and Y to cover cool red and brown dwarf stars, C to cover stars that are rich in carbon and D to cover white dwarf stars.

Stellar Spectra

The light we receive from the stars is rich in information. This can be extracted by looking at the spectrum of the star, essentially taking its light and spreading it out to reveal its constituent colours. When this is done, a rainbow spectrum of colour is revealed that is crossed by dark, so-called absorption lines. These lines reveal the chemical fingerprint of the star – which atoms are present and their relative abundance.

It's the pattern of lines in a star's spectrum that ultimately gives it its place in the spectral classification scheme and the multitude of different "fingerprints" discovered, which has created such an extensive range of class-types.

The spectrum of a star can be obtained by passing its light through a diffraction grating, an optical device that splits light by the physical process of diffraction. In this case incoming light encounters a regular pattern of parallel rulings on an optical surface. Gratings may work by allowing light to pass through them (transmission grating) or by reflecting light off a surface (reflective grating). A simple yet effective way to see the effects of a reflective grating is to reflect light in a CD or DVD disc. If the disc is held in a certain way, the light is seen spread into a spectrum of colours.

In addition to being able to tell us the chemical make up of a star, spectral analysis can also be used to measure magnetic field strength by monitoring how spectral lines appear to split due to a phenomenon known as the Zeeman Effect. Shifts in the position of a star's spectrum can also indicate the speed a star is moving relative to the Earth, due to the Doppler Effect. It's this effect that can be used to determine the presence of

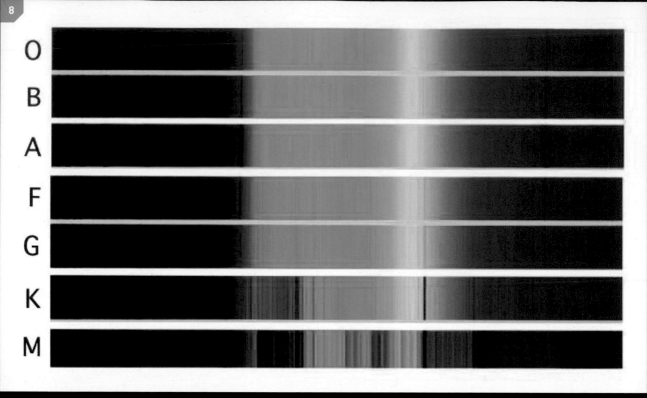

O
B
A
F
G
K
M

pairs of stars too close to be separated visually. As the stars orbit each another, the two sets of spectral lines can be seen to undergo a periodic shift relative to one another. This shows what visually looks like a single star, to be a pair. Such a system is known as a spectroscopic binary.

Double and Binary Stars

On the subject of paired stars, gravitationally bound stars that orbit one another are abundant in our Galaxy. In fact our Sun is unusual in being alone. A telescope shows many examples of multiple star systems in the night sky, some of which simply look as if they are close by a line of sight effect. In reality, they could be many billions of miles apart and not connected at all. Such arrangements are known as optical double stars. In addition, some stars are gravitationally bound to one another often taking many years to complete their orbital dance. Such systems are called binary stars. One of the most beautiful binary star systems visible in the northern hemisphere is that known as Albireo, which marks the beak of Cygnus the Swan. Seen with a small telescope this seemingly single star, splits into a yellow primary with a bluish companion close by. It's not known whether this is a true binary system, but if it is, both components will take in the order of 100,000 years to complete one orbit.

Mulitple star systems abound in the night sky, with binaries, triples, quadruples, quintuples, sextuples and even a possible septuple, a seven-star system known as Nu Scorpii.

Variable Stars

We have already talked about double stars, binary systems and star colours. Even a small telescope will show plenty of examples of what we mean. In addition, many stars show variation in light output and these are what we refer to as variable stars. The variation may come about because of external or internal processes. An example of an external process is the so-called eclipsing binary. One of the best examples of this type of star is Algol in the constellation of Perseus. Also known as the "Winking Demon" or Beta Persei, to the naked eye Algol dips in brightness every 2 days 20 hours and 49 minutes, remaining dim for around 10 hours. The dimming is due to the fact that Algol is a binary star system. One component is bright, the other dim and from Earth one star appears to cross the face of the other, partially covering its light from view. When the bright star covers the dim one, the dip in brightness is only detectable using specialist equipment. When the dim one covers the bright one however, the dip is appreciable and it's this which makes the Demon "wink". There are many examples of eclipsing binary stars.

Other stars vary because of internal processes. Old, large red stars for example, vary because they are running out of fuel. This leads to instability and changes in luminosity. One of the finest examples is the star Omicron Ceti, also known as Mira, in the constellation of Cetus, the Whale. This remarkable star varies in brightness over a period of 332 days. At its brightest, it can easily be seen with the naked

[8] Dark absorption lines provide spectral finger prints for stellar classification.

[9] A misty sky helps bring out the colours of stars in the constellation of Orion, the Hunter.

eye as an addition to the constellation, rivalling the brightest member of Cetus, a star known as Menkar. However, at its dimmest, Mira disappears from view, dropping down to the limits of binocular visibility. No wonder then that the name Mira literally means "wonderful".

Another example is Mu Cephei in the far north of the sky, a star also known as Herschel's Garnet Star. Mu varies between magnitude 4 and 6 so it is generally a naked eye object and always within binocular range. Use a telescope and it appears like a glowing coal because its temperature is so much lower than our own Sun. To compensate Mu Cephei is immense and would fill our Solar System out as far as Saturn. It will explode as a supernova this year, next year, or a hundred years, or a million years hence – we cannot know.

We can have far brighter examples of star colours. In the main winter constellation of Orion, the two leading stars are Betelgeux, an orange red supergiant, and Rigel, a very bright powerful blue star well over 40 thousand times more luminous than our Sun. Both these giants evolve much more quickly than our Sun, though of course one cannot see any detectable change over periods of a good many lifetimes.

To find the various objects mentioned here it is merely a question of time and patience, using the star maps at the back of this book. In addition to the "fixed" stars of the night sky, there can be unexpected visitors like novae which appear or "new" stars.

In fact a nova is not a new star at all but the white dwarf companion of a normal star. The white dwarf pulls material away from its companion, increases its brightness and then

fades back to obscurity. Some novae may be very brilliant. Nova Aquilae in 1918 surpassed every star in the sky apart from Sirius, but did not last long and has now become very faint indeed.

Clusters and Nebulae

Stars form out of collapsing clouds of predominantly hydrogen gas. Huge clouds of gas may collapse to form many stars which are all moving through space in the same direction and at the same speed; a quantity known as proper motion. Stars formed in this manner form what is known as an open cluster and there are many examples of this type of association in the night sky. Probably one of the most famous open clusters known is the Pleaides or Seven Sisters, located in the constellation of Taurus, the Bull.

Best seen in the autumn and winter skies, to the naked eye, the Pleiades shine like brilliant diamonds scattered on black velvet. Despite being called the Seven Sisters, those with keen sight may be able to make out many additional members of the cluster. The author Pete Lawrence has seen 15 under good conditions from his home in Selsey, England. Binoculars or a small telescope on a low power will show many further members, the cluster containing around 250 stars in total. A long exposure photograph of the Pleiades shows the stars to be surrounded by delicate blue gas. This was once thought to be left over material from the formation of the stars but the current belief is that Pleiades stars, moving through our galaxy, have simply encountered the cloud of material. The light from the

9

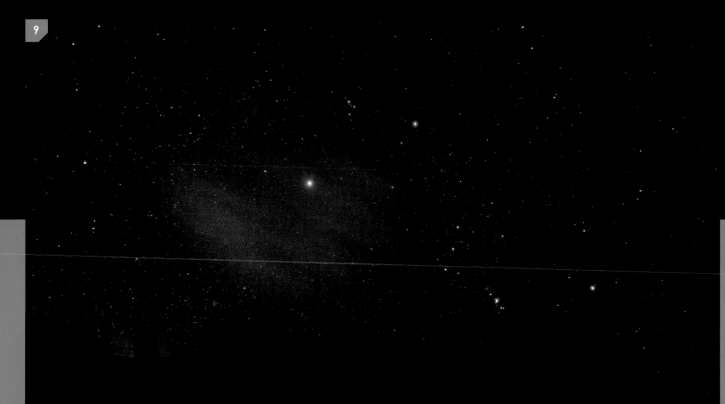

[10] The wonderful Cocoon Nebula lies in the northern regions of Cygnus, the Swan. This object consists of a cluster of stars known as IC 5146 and a region which exhibits both emission and reflection characteristics known as Sharpless 2-125 (Sh2-125). Wide field shots of the nebula show it sitting at the end of a long finger of darkness caused by gas and dust blocking the background stars of the Milky Way Galaxy.

[11] A long exposure of the Pleiades cluster (Messier 45) reveals a delicate reflection nebula – galactic dust illuminated by the light of the cluster stars. (Ian Sharp)

[12] Glowing gas and clusters make up the region of sky known as Orion's Sword. At the top lies a beautiful reflection nebula known as NGC 1977. Dark lanes of dust crossing the nebula give the appearance of a running figure and the nebula is known informally as the "Running Man Nebula". Below NGC 1977 lies the bright gas cloud known as the Orion Nebula or M42. The comma-shaped region to the upper-left of the brightest part of the nebula is classified separately from the main nebula and is known as M43. Finally, the sword tip is defined by a small open cluster below M42 known as NGC 1980.

stars reflecting off the dust in the cloud gives the nebula its blue appearance.

The two main types of nebula (Latin for cloud) are emission and reflection. Reflection nebulae, such as that which permeates the Pleiades open cluster, tend to be bluish in colour while emission nebulae have a reddish component to them. There are many superb examples of emission nebulae visible in the night sky, including the Eta Carina Nebula, Tarantula Nebula, Orion Nebula, Lagoon Nebula and the list goes on. Many of these beautiful objects are within reach of amateur equipment and many make superb photographic targets. The red colouration mainly comes from light emitted by hydrogen atoms, the clouds of gas being predominantly made of hydrogen. The hot young stars forming within the cloud ionize the atoms of hydrogen and cause them to

give off specific wavelengths of light. One of the strongest wavelengths emitted is that called hydrogen alpha (656.28nm) and this lies in the red part of the spectrum.

The bright Orion Nebula which lies at the heart of Orion's sword, is visible to the naked eye, one of only a few. At the nebula's heart lies a hot young cluster of stars known as the Trapezium Cluster, stars which have literally been born out of the nebula material.

Another interesting type of object occurs when a dark cloud sits in front of a brighter one. One of the best examples of this is the Horsehead Nebula, again located in the constellation of Orion. Here a finger of dark gas protrudes in front of a curtain of emission nebulosity. The finger looks like a knight chess piece in silhouette and this gives the nebula its distinctive name.

Open clusters and nebulae provide many hours of

44), also known as Praesepe (Manger), lies at the heart of Cancer.

enjoyment for astronomers. Hunting them down and revelling in their appearance is one of the great joys of amateur astronomy. With one or two notable exceptions, these objects lie within our own Milky Way Galaxy. Scattered in a halo around the core of the Milky Way is another type of cluster, known as a globular cluster. Here, tens of thousands or even millions of stars are clustered together in a tight region of space. Each star is in orbit around the common centre of gravity of the cluster but collisions are very rare. There are many great examples of globular clusters within amateur reach, including the Great Globular in Hercules which is visible in the Northern Hemisphere. It has to be said though, that the Southern Hemisphere has the best examples with objects such as Omega Centauri and 47 Tucanae almost literally having the ability to take your breath away when viewed through a telescope.

Cataloguing the Universe

There are many catalogues of these objects in existence. One of the most famous is the Messier Catalogue, compiled by the French astronomer, Charles Messier in 1771. The list was compiled to aid Messier in his hunt for comets, creating a list of objects that he didn't want to keep tripping over. Today the 110 objects which now reside in the slightly extended original catalogue, form essential viewing for amateurs the world over with objects such as Messier 45, the Pleiades, and Messier 42, the Orion Nebula. The entries are normally simply listed with an M prefix.

The Messier catalogue is biased towards the Northern Hemisphere although many of its objects can be seen from the Southern Hemisphere too. Patrick created his own list which he called the Caldwell Catalogue. In truth this wasn't meant to be taken seriously but has stuck. The Caldwell Catalogue balances the hemispheres somewhat by starting

in the far north with C1, an ancient open cluster in the constellation of Cepheus, and progressively moving further south with each new entry. The last entry, C109 is a planetary nebula in the southern constellation of Chamaeleon, located roughly 9° from the South Celestial Pole.

The more extensive New General Catalogue (NGC) contains 7,840 objects boosted by two further Index Catalogues (IC) adding an additional 5,386 objects.

The catalogues provide information about the object's appearance and position in the night sky. Positions are given by a coordinate system known as Right Ascension and Declination (RA & Dec.). Declination indicates the angular distance of the star from the celestial equator and is equivalent to latitude on the surface of the Earth. Right Ascension requires a little more explaining, but corresponds more or less to celestial longitude. Once these coordinates are known, if you are using a Go-To telescope, it is merely a question of punching in the coordinates and letting the telescope do the rest. If you don't use Go-To, the objects are best located by finding them on a chart and star hopping to their location.

Observations with the naked eye are very satisfying but today many astronomers rely on photography.

Photographing the Stars

Photographing the stars, clusters, nebulae and other galactic wonders that exist within our own galaxy is something that can be undertaken with quite modest equipment. At its simplest, setting a camera on a tripod and pointing it up to the night sky with the shutter open for several minutes should record the stars as trails. The motion that causes this isn't anything to do with the stars at all of course, but rather by the fact that the Earth on which you and your camera are standing, is rotating underneath them. All of the stars and

hemisphere, the north celestial pole is almost marked by the star Polaris. In the Southern Hemisphere the closest naked star to its celestial pole is the rather dim Sigma Octantis. Sigma lies over a degree (two apparent full Moon diameters) away from the actual pole and is at the limit of naked eye visibility.

If you point a fixed camera at one of the celestial poles and take an extended shot of the stars, they will trail around the pole, tracing out arcs of circles. If you could expose the shot for a whole day, the arcs would eventually join up and create complete circles.

Photographing the stars and other galactic objects as static objects is more complex because here, the more magnification you use, the more the apparent motion caused by the Earth's rotation is magnified too. If you use a wide angle lens on a fixed tripod mounted DSLR camera for example, a short exposure of less than 30 seconds will typically show little in the way of star trailing. If the power is increased by using a telephoto lens or even coupling the camera to a telescope, the same length of exposure taken from a fixed mount will definitely show trailing.

Unfortunately, many deep sky objects require long

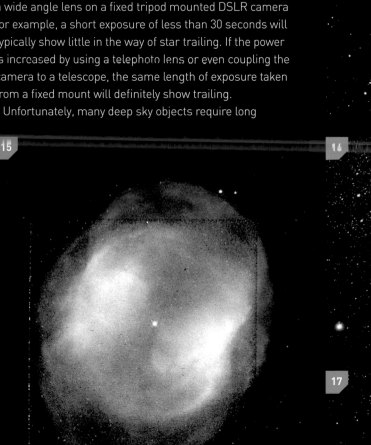

[15] C109 (NGC 3195) the last object in the Caldwell catalogue. It is a planetary nebula. This image is from the Hubble Space Telescope.

[16] Star clusters and nebulosity abound in the North America Nebula (NGC 7000) located in Cygnus, the Swan.

[17] A chance stellar alignment produces an asterism (unofficially recognised pattern) that, in this case, resembles a coathanger (Brocchi's Cluster).

exposures in order to capture them well, and to do this it's necessary to allow for the rotation of the Earth, and compensate for it. The most common way to do this is to use a camera and/or telescope mounted on a driven equatorial mount. This is a mount with one axis set parallel to the rotational axis of the Earth so that the motion of a telescope fixed to the mount follows the natural motion of the stars and planets across the sky. If the mount is driven at the same rate, but in the opposite direction to the rotation of the Earth, the object that's being looked at through the telescope no longer appears to move. If the mount is set up accurately, it's possible to take extended exposures of the stars using a camera attached to the telescope. However, even here there are limits and exposures greater than a couple of minutes may still start to show small trailing effects. These occur due to small errors in the mount's pointing accuracy and minor variations in the telescope mount's drive rate.

[18] A magnified composite showing 24 hours of star rotation around the north celestial pole. The brightest ring in the shot is caused by Polaris, the "North Star". Note how many fainter stars there are between Polaris and the pole.

[19] An inexpensive way of taking a flat field is to cover the end of the imaging telescope with a taut white cloth and take a photo with the telescope pointing at an evenly lit patch of sky. It's important that the camera and telescope stay in the same configuration between imaging and taking the flat.

[20] Hot pixels (arrowed) are a form of static or predictable noise in an image. By taking an exposure of a similar length with the telescope's cover fitted, the resulting dark frame image contains only static noise. Subtracting this from the original light frame removes the static noise.

For longer exposures, the use of an autoguider is recommended. This is a more sophisticated setup that uses a second telescope physically attached to the first with a camera dedicated to monitoring where the second telescope is pointing. This autoguiding camera is, in turn, controlled by autoguiding software. The basic idea here is that you take an image through the autoguiding telescope, pick a bright star from the image and then instruct the autoguiding software to keep that star in the same position in the image field. In order for this to work, the computer running the autoguiding software needs to be physically connected to the mount. Many mid- to high-end mounts currently offer this facility.

Autoguiding is a lot easier than it used to be but it can still require quite a bit of setting up. The connections necessary to make it work, including the various power cables needed, can turn the business end of a telescope into a veritable bird's nest of wires if you are not careful. It's also very important that the scope and its various attachments are properly balanced on the mount.

A properly set-up autoguiding system will allow you to take exposures lasting theoretically for as long as you like as long as the guide star is above the horizon and visible. However, practicalities do apply here too and in reality, it's unusual for exposures to go beyond a maximum of 30 minutes.

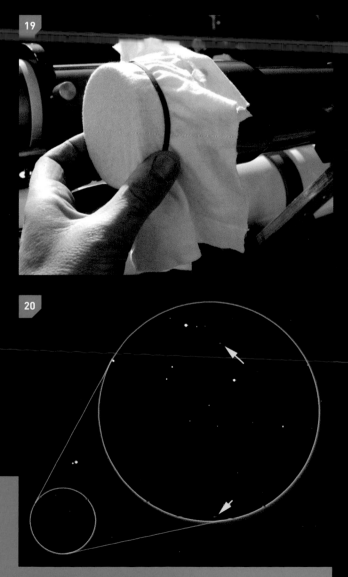

Noise

Digital imaging has revolutionised astrophotography in many ways but not everything in the digital garden is rosy. One major issue with digital imaging systems is noise that manifests itself as stuff in your images which really should not be there. There are many different causes of noise but the types break down into one of two categories: random noise and static noise.

Static noise appears as a fixed and more importantly, predictable addition to an image. An example of static noise would be a pixel in your camera sensor that's permanently in the on state, known as a "hot pixel". The simplest way to deal with static noise in images is to follow the normal exposure(s) with a similar length of exposure with the lens cap fitted. This creates a "dark frame" which records just the static noise in an image. The dark frame image can then be subtracted from the normal exposure (known as a "light frame") correcting the static noise issues. It's important to take darks before, after or even in-between normal exposures so the temperature of capture is the same. The camera settings should also be kept constant.

Random noise is less easy to deal with and unfortunately also affects dark frames as well as light frames. It has numerous sources, and is especially affected by heat which causes false signals to be recorded across an imaging sensor array. One obvious way to reduce heat related effects is to cool the imaging chip, and this is exactly what happens in a cooled astronomical CCD camera. However, even here, there will be a random pattern of very weak signals lurking in the background of the image.

Part of the process of massaging the image into a visible form is to stretch the data within the image file – basically making the brightest part of a faint image as bright as it will go while keeping the dark parts close to black. If there is random noise lurking in the shadows of an image, stretching the image will increase its visibility too. The way to deal with random noise is to take lots of exposures and average them together. This strengthens the appearance of the subject while helping to smooth out the random background noise.

Optical noise occurs because of defects in the optical path leading up to the camera. This is typically caused by dust and obstructions to the light path. A commonly encountered problem occurs when the light cone coming from the optics is clipped by the edge of the eyepiece holder. This produces a graduated darkening towards the edge of the frame known as vignetting.

The effect of optical noise can be removed from the light frames by taking a flat field image, also known as a "flat". A flat is taken by exposing the telescope to an evenly illuminated light source and taking an underexposed (approximately to half saturation) shot of the view. This is then divided into the light frame, removing the optical noise in the process.

The process of applying all of these corrections is called calibration and is a necessary evil if you want the best results when photographing deep sky objects. Although it sounds complex, the process is not too difficult once you get into a routine. The degree to which you're prepared to go in terms of calibration is a personal choice and it's certainly possible to get impressive results without going through all the stages.

However, if you do apply calibration, the results will typically be much better.

A typical deep sky imaging session consists of the following stages:

1) Set up
2) Capture
3) Calibration
4) Assembly
5) Image processing

1) The setup stage is self explanatory and covers the process of setting your telescope up, finding the target, framing the target and focusing.

2) Capture is the process of taking the light frames, dark frames, flats and other calibration files should you wish to use them. This stage typically involves taking a number of light frames of the same exposure length. These are sometimes referred to as sub-frames and the greater the number taken, the more random noise reduction can be applied. Random noise reduces as the square root of the number of frames captured. For example, if you take 4 light frames, averaging them will reduce the amount of noise by a factor of 2; the square root of 4. Take 9 light frames and averaging them reduces the random noise by a factor of 3 and so on.

Following the capture of the light frames, the telescope should be capped and a number of dark frames taken of the same exposure length as that used for the lights. These will contain their own random noise so once again, these need to be averaged to produce what's known as a "master dark". The more dark frames you average, the more the random noise that they contain will be smoothed out.

The flats are trickier to achieve because in order to take them you need to be able to point your scope at a uniform light source. A simple way to achieve this is to wait until morning and cover the end of the scope with a taught bit of white cloth, for example from a white tee-shirt. Point the scope at an evenly illuminated patch of cloudless sky and take a photo. The photo needs to be between a third and a half exposed for the best results. Once again, flats also contain random noise so take a number of them and average them together to produce a "master flat".

Alternative methods to take flat fields include building your own light box or using an electroluminescent panel which can be held over the front of the scope to give a uniform plane of illumination. It's very important that the camera's position relative to the telescope is not changed between taking the light frames and the flats. If this does occur, the flats will be useless. A further type of calibration image is the bias frame. This is quite easy to take and simply requires you to put the lens cap on the telescope and take the shortest exposure possible.

3) Once all of the calibration files have been successfully captured, the next step is to apply them. There are various commercial applications which can do this automatically as long as you tell the program which file is which. Alternatively,

there are freeware options available which will do a similar job. One popular freeware application to perform all calibration steps automatically as well as assembling the final calibrated files into a finished image is DeepSkyStacker (http://deepskystacker.free.fr/english/index.html).

4) Once you've created the end result, you may feel a little cheated as it may not look that great to start with. This is where the final "image processing" stage is applied and is the one that will either make or break the final image. There are many dedicated programs available to handle astronomical image processing and much of the subject is beyond the scope of this book. A layer based graphics program such as PhotoShop, Paint Shop Pro or the freeware GIMP is useful for tweaking the final image but for the best results a program that can perform what's commonly known as a DDP or Digital Darkroom Process on the image will be the best place to start.

Amongst some of the commercial applications that can do this are Maxim DL, AstroArt and ImagesPlus. The powerful, yet sometimes daunting freeware IRIS package is also well worth considering. The DDP is basically used to stretch the image data but designed to bring out the fainter detail without causing the brighter detail to appear over-exposed.

If all this seems like too much hard work from the description given, don't worry, you wouldn't be the first to think that. If this is how you feel then start out by taking some basic images of bright subjects and progress from there. Remember no one's judging your efforts and the experience should be an enjoyable one. Your first astrophoto taken through a telescope will give you a thrill, irrespective of its appearance.

More advanced imaging can be carried out by using mono cooled CCD cameras and using filters. Of particular interest are narrowband filters which restrict the light coming into the camera to specific wavelength bands. Examples of astronomical narrowband filters are those which centre on the H-alpha, OIII, H-beta and SII. Imaging through these filters allows you to concentrate on the light given off by specific atoms in the target object. As such, the information captured can show where these atoms are concentrated.

An additional bonus for using certain types of narrowband filter is that many are tuned to regions of the spectrum which are not emitted by streetlights. As an example, imaging with H-alpha light in badly light-polluted skies can still produce amazing results.

Planets Around Other Stars

Our Solar System is not unique in the Universe and advanced techniques to micro-analyse stars are revealing the existence of other worlds, referred to as exoplanets, in orbit around distant suns. An area currently the realm for professional or advanced amateurs, it's an amazing journey of the mind to imagine these alien worlds of all shapes and sizes scattered throughout our Galaxy and no doubt the Universe.

Known as exoplanets, the count of discoveries increases daily thanks mainly due to a spacecraft called Kepler which monitors a region of sky roughly 12 degrees in diameter in the constellation of Cygnus, the Swan. Here it looks at the light of 150,000 stars looking for tiny dips in output caused by planets passing across the star's disc. For a terrestrial sized planet, the dip would typically be around 100 parts per million of the star's light. When three such transits have been seen and confirmed, then the object is certified as a valid exoplanet. At the time of writing there are in excess of 700 certified exoplanets known and thousands waiting for verification.

Of course, the holy grail of planet hunting would be to locate a world much like our own which exists at a distance from its parent star (the Goldilocks Zone) where water can be liquid on its surface. At the time of writing numerous "Super Earth" candidates have been put forward: planets larger than Earth but located within or close to the habitable zone, such as Kepler 22b. As our techniques and equipment improves, no doubt we will discover planets similar to Earth but the question as to whether they will harbour life still remains unanswered. Probability and the history of discovery would tend to suggest that the finding of extraterrestrial life in some form is very likely.

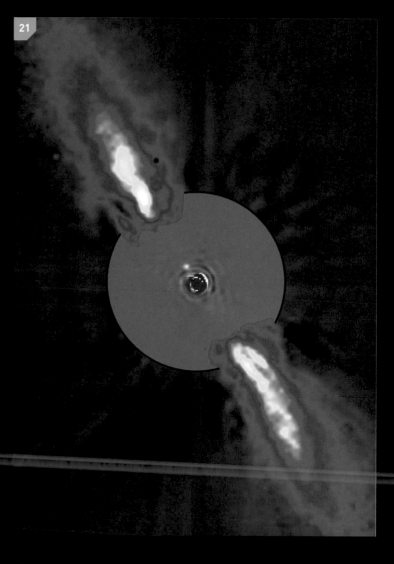

[21] Beta Pictoris b, the extrasolar planet can be seen as the white dot just outside the central dark blue circle, above and to the left. The star Beta Pictoris itself lies at the centre of the blue circle. The lobes are dust, and the image is captured at infrared wavelengths. ESO, A.-M. Lagrange et al.

[22] Fomalhaut is a young star 25 light years away. In this composite image from the Hubble Space Telescope, the star is at centre, and a planet about three times the size of Jupiter has been identified orbiting Fomalhaut at a distance of 10.7 billion miles. Known as Fomalhaut b, it is shown larger in the detail picture at bottom right. (NASA/ESA/University of California, Berkely/JPL/Goddard/Lawrence Livermore Laboratory.)

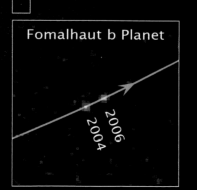

Fomalhaut b Planet

2006
2004

BEYOND
THE STARS

This picture of the spiral galaxy NGC 247 was taken at
ESO's La Silla Observatory in Chile. NGC 247 is thought to
lie about 11 million light years away in the constellation of
Cetus, the Whale.

CHAPTER 9

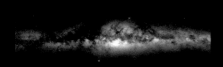

So far we have confined ourselves to our own Galaxy. As recently as the 1930s most of the cloud-like objects in the heavens were referred to as nebulae (Latin for clouds). Some nebulae are indeed clouds of gas in which fresh stars are being born. Others though are quite different – they are huge island cities of stars called galaxies, which exist outside of our own Milky Way.

M31, the giant galaxy in Andromeda, looks like a mass of stars, and was previously known as a "starry nebula". It was thought to lie within our Milky Way Galaxy, but is in fact an external galaxy in its own right, 2.5 million light years away. It is the most remote object that can be seen clearly with the naked eye. It was not until 1930 that Edwin Hubble, using the most powerful telescope in the world, realized that this object could not be inside our own Galaxy.

He did this by identifying so-called Cepheid variable stars in various "starry nebulae", including Andromeda. Cepheid variables have periods of several days or several weeks, and we always know how a Cepheid is going to behave. In 1930 it was discovered that the real luminosity of a Cepheid depends upon its period: the longer the period, the more powerful the star. As soon as we can indentify a Cepheid and measure its period we know how luminous it really is and this gives us a good estimate of its distance, since there is a straightforward relationship between distance and luminosity. Cepheids are very powerful so can be identified out to great distances. We refer to them as "standard candles" in space.

We now know the number of galaxies in our range is very great, each containing huge numbers of stars, like our Milky Way. An average galaxy probably contains several hundred thousand million stars. Astonishing as this figure is, current estimates put the number of galaxies that may exist at from several hundred thousand million, perhaps even up to a million million (a trillion).

Galaxies come in a variety of shapes and sizes and many of these can be seen in the night sky with quite modest equipment. Our own Milky Way Galaxy is thought to be what is known as a barred spiral – that's a spiral galaxy with a star-dense core that appears elongated along an axis.

1

[1] The Markarian Chain of galaxies in the Virgo Cluster.

[2] Hubble's tuning fork classification for galaxies: the main characteristic shapes are Sa, Sb, Sc along the upper parallel branch, and on the parallel branch below Sba, SBb and SBc

[3] The Sombrero Galaxy (Messier 104) in Virgo.

One of the best galaxy hunting grounds in the entire night sky is in a region of sky known as The Realm of Galaxies. This lies within a large arc of stars in Virgo, the Virgin, often referred to as the "Bowl of Virgo". The region bounded by the five stars that make up the bowl – Epsilon, Delta, Gamma, Eta and Beta Virginis – and the star Beta Leonis, or Denebola, marks the direction to two large clusters of galaxies known as the Virgo Cluster and Coma Cluster. Just as stars are found gravitationally bound into open clusters, galaxies too are found in gravitationally bound groups.

A telescope on a low power will show many stars in the region that appear slightly fuzzy. These are individual galaxies and here you'll find a large collection of all types including spiral, lenticular and elliptical. There may be as many as 2000 members in the Virgo cluster alone.

Galaxy Classification

The shape of galaxies is often represented by what's known as the Hubble Tuning Fork Diagram, after the famous astronomer Edwin Hubble. The diagram depicts the progression of different observed galaxy shapes, from elliptical through to spiral. Spiral galaxies are arranged in two parallel branches, one for normal spirals and one for barred spirals. The diagram is normally arranged so that the shape of the galaxy types gets more extreme towards the right side of the diagram. The ellipticals are arranged in order of eccentricity, or how elliptical they appear. They are identified by the letter E followed by an index number.

The shapes range from E0, which is spherical, through to E7, which is highly elliptical. Beyond this point the flatness of the galaxy cannot be sustained. The most common type of elliptical galaxy is E3. The point where the "handle" meets the "forks" is

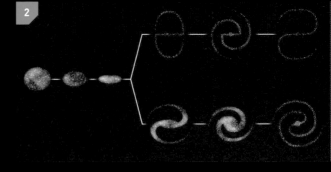

where a special class of galaxy known as a "lenticular" sits. This is a flattened galaxy type similar in profile to a spiral except that it has no spiral structure in the disc of stars that surrounds the core. Lenticulars are given the designation S0. From here the branches of spirals, identified by the letter S, and barred spirals, identified by SB, begin.

Both spiral types are further divided into sub-types a, b and c, which indicate the tightness of their spiral arms. Type Sa and SBa galaxies are tightly wound while Sc and SBc have very open arms. The spiral arms of galaxies are regions where material is being compressed and stars are being born. The arms often look bluish in colour due to the hot, young blue stars which are forming there. The core region, by contrast, often looks yellow due to the large population of older, redder stars that can be found there.

The Hubble Tuning Fork Diagram isn't perfect and we now know of several types of galaxy which the tuning fork is unable to classify properly. The De Vaucouleurs system is designed to address this shortcoming by adding another fork to the diagram for weakly barred spirals. The De Vaucouleurs system is also able to classify galaxies surrounded by a ring of stars, compact galaxies, dwarf galaxies, peculiar galaxies and galaxies with active galactic nuclei.

To the Edge of the Universe

By examining the spectra of galaxies, we can find out how fast they are moving. If the dark lines in their spectra are shifted towards the end of the spectrum it means the light source is receding. The further away a galaxy is, the greater the redshift, and the faster it is moving away from us. All the galaxies are receding from each other. Our telescopes allow us to see nearly 14,000 million light years into space.

There is one major unexplained fact. If the Universe – space, time, everything – began at one instant known as the Big Bang (a term used in a derogatory sense by UK astronomer Fred Hoyle, who never believed in it), when the galaxies began expanding, gravitational pull would slow them down. Instead,

the rate of expansion is increasing. Astronomers talk about dark energy as being responsible for the acceleration of the expansion, but no one really knows what it is.

How far can we go? Beyond a certain distance, the rate of expansion becomes equal to the speed of light, with all manner of complications; we cannot see that far, so do not have an idea of what goes on in these remote parts of the Universe.

How big is the Universe?... We do not know! No doubt there are large parts of it we cannot see, and there is an important point: either the Universe is infinite or not. If it is finite, what lies outside it? Our brains are quite unequal to this task. We are still woefully ignorant about fundamentals. Energetic efforts are being made to search for signs of life in other solar systems. In view of the immense distances involved, it is hardly surprising that we haven't discovered any yet. Our most promising line of investigation involves radio waves but even this has not helped. There may well be other beings in the Universe more advanced than us who can.

We cannot travel to other solar systems by any means known to us. Rockets would take many centuries, unless we have a major breakthrough. But one never knows, Science fiction does tend to turn into science fact and in future we may discover how to tackle it. Whether we may meet another race remains to be seen. Time will tell.

Photographing Galaxies

The process of photographing galaxies is similar in essence to that described in Chapter 8 for photographing the stars. The main issue that needs to be taken into consideration is that most galaxies are quite faint. Even the mighty Andromeda

[4] Pinwheel Galaxy (Messier 33), a spiral galaxy in Triangulum, photographed by Ian Sharp with an 80mm refractor and a cooled CCD camera.

[5] Bode's Galaxy (Messier 81) in Ursa Major by Ian Sharp. Telescope was a Vixen VC200L with a cooled CCD camera.

[6] The Hubble Space Telescope's Ulta Deep Field image at highest resolution reveals around 10,000 distant galaxies, some dating from 400 million years after the Big Bang.

Galaxy, M31, is misleading in this respect because the bright elongated smudge that you can see with the naked eye, binoculars or a low-power telescope is just the core of the galaxy. The real detail lies in the outer extremities of the object. Here you'll find the spiral arms littered with bunches of star clouds.

To capture everything the galaxy has to offer, it's necessary to take a long exposure to capture the outer arms. This can sometimes lead to over-exposure of a brightly lit core region and one technique to overcome this is to take two sets of images – one for the faint outer regions and one for the core. Both shots need to be taken following the same calibration process as described in the previous chapter. When done, you should end up with two images of the same object. One will show a bright core with barely any spiral arms visible. The other will show the spiral arms well but the core will be washed-out.

Both images should then be loaded into a layer-based graphics editor such as Photoshop so that the over-exposed-core version is on the top of the stack. It's important at this point to make sure both layers are aligned together in respect of the stars they contain. This can be done using the move tool and using the keyboard's cursor controls to gently nudge the upper layer's position. Toggling the visibility of the upper layer off and on will reveal any misalignments. Rotational misalignments can be avoided by making sure that both sets of images are taken during the same session without moving the camera in between.

Once aligned, make sure the upper layer is visible and use a magic wand tool to select the bright core. It's not that crucial to make sure the selection is right to the very edge of the core – the selection can be approximate. The next thing to do is to create what's known as a layer mask for the upper layer. In Photoshop this is done by first copying the selection to the clipboard and then holding down the Alt key (on a PC or some Macs) or option key (on a Mac) while pressing the layer mask button shown below the list of layers (normally a grey rectangle with a hole in the middle).

This creates a layer mask with the copied selection. A layer mask is essentially a map of what should and shouldn't be shown from the layer being masked. Anything which is white in the mask itself indicates that the layer should be opaque. Anything which is black in the mask indicates that the layer should be transparent. Shades of grey indicate partial transparency – the darker the grey the greater the amount of transparency applied.

The layer mask created will show the properly exposed core from the bottom layer through the transparent hole that's been created in the upper layer. The hole will, at this stage, have sharp edges and look rather awkward. The magic happens by selecting the layer mask and applying a Gaussian blur to it – a common blurring function found in most graphics editors.

The degree of blur applied should be adjusted to allow the two layers to merge naturally. Once done, it's possible to adjust the appearance of the lower, properly exposed core independently of the upper, properly exposed arms.

Photographing Galaxy Clusters

The process required to capture images of faint galaxy clusters is no different to that described in Chapter 8 for deep-sky photography. These distant and consequently rather faint objects may require long exposures of anywhere between 5 and 30 minutes to capture well so an autoguiding set-up is highly recommended. The delicate streams of starlight that can occur between galaxies which are undergoing mutual gravitational disruption can be fascinating to capture but do tend to require longer focal lengths to provide enough image scale to pick them up convincingly.

Unlike when imaging the planets, the faintness of these objects means that it's necessary to grasp as much light as possible. Consequently, large apertures will fare better than smaller ones. However, where clusters of galaxies are concerned, one of their most fascinating aspects is the shells-on-a-seashore appearance all of the different galaxies can take on.

Where there's a mix of face-on spirals, edge-on spirals and ellipticals, a wide field capture that shows the shape of each member can make for an amazing group photograph. Here, the equipment requirements are less demanding and such shots are well suited to small wide-field refracting telescopes. Again, the relative faintness of the objects does warrant longer exposure though, so autoguiding is once again recommended.

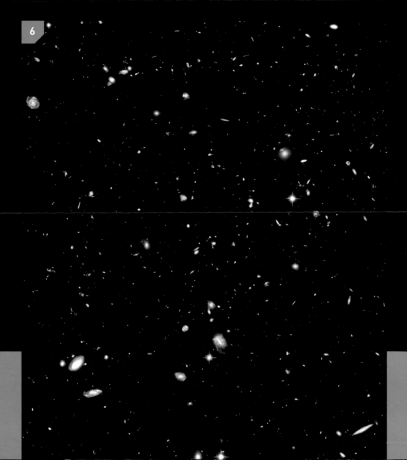

6

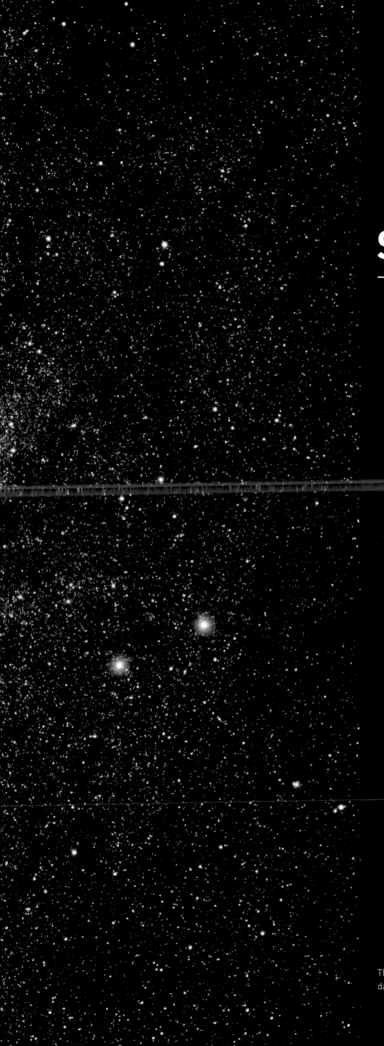

STAR ATLAS

The four bright stars of the Southern Cross shine brightly to the left of the dark Coalsack Nebula. Credit ESO/Yuri Beletsky.

USING THE STAR MAPS

In the Star Atlas the stars are mapped month by month, first for the Northern Hemisphere, then for the Southern. For each month the stars you can see are shown for an observer facing north or facing south. If you hold the map for a particular month up in front of you, while facing either north or south, then you will be able to identify the stars that are visible to the naked eye.

FINDING THE PLANETS
Observing Mars, Jupiter and Saturn

The positions of the planets visible to the naked eye that orbit the Sun further out than the Earth are marked on the relevant charts. As these positions change over time, the positions marked indicate where the planets will be when at opposition for the year indicated.

Opposition

Opposition is a term used to describe when a planet lies opposite the Sun in the sky, and appears at its brightest. At this time, a planet will be visible all night, rising around sunset and setting around dawn.

Of the planets marked, Mars moves quite rapidly across the sky before and after opposition while Jupiter and Saturn don't.

Observing Mercury and Venus

The inner planets, Mercury and Venus, never wander too far from the Sun and are best seen when they are at elongation.

Elongation

Elongation is a point in the orbits of Mercury and Venus when the planet will appear at its greatest distance from the Sun. In order to determine the best time to see the two inner planets, we've listed their elongation dates in tables. Eastern elongation means the planet is visible in the western part of the sky after sunset, while western elongation means it's a morning object, rising before the Sun.

Mercury can never appear to wander from the Sun by more than 28 degrees while Venus can fare better with a maximum elongation of 47 degrees.

Mercury in the clouds in the twilight skies.

ELONGATIONS OF MERCURY

Date		Elongation	Type
2015	24 Jun	17:07 22° 28' 47"	West
2015	4 Sep	10:19 27° 08' 12"	East
2015	16 Oct	03:16 18° 07' 25"	West
2015	29 Dec	03:11 19° 43' 14"	East
2016	7 Feb	01:23 25° 33' 02"	West
2016	18 Apr	13:59 19° 55' 33"	East
2016	15 Jun	08:45 24° 10' 46"	West
2016	16 Aug	21:21 27° 25' 58"	East
2016	28 Sep	19:27 17° 52' 31"	West
2016	11 Dec	04:38 20° 46' 05"	East
2017	19 Jan	09:43 24° 07' 57"	West
2017	1 Apr	10:17 18° 59' 41"	East
2017	17 May	23:23 25° 46' 51"	West
2017	30 Jul	04:38 27° 12' 07"	East
2017	12 Sep	10:16 17° 55' 59"	West
2017	24 Nov	00:26 21° 59' 32"	East
2018	1 Jan	19:58 22° 39' 32"	West
2018	15 Mar	15:09 18° 23' 59"	East
2018	29 Apr	18:23 27° 01' 20"	West
2018	12 Jul	05:29 26° 25' 06"	East
2018	26 Aug	20:34 18° 18' 59"	West
2018	6 Nov	15:31 23° 18' 56"	East
2018	15 Dec	11:29 21° 16' 12"	West
2019	27 Feb	01:24 18° 08' 14"	East
2019	11 Apr	19:41 27° 42' 46"	West
2019	23 Jun	23:15 25° 09' 23"	East
2019	9 Aug	23:07 19° 02' 47"	West
2019	20 Oct	04:01 24° 38' 00"	East
2020	10 Nov	17:03 19° 05' 52"	West
2021	24 Jan	01:58 18° 33' 36"	East
2021	6 Mar	11:23 27° 15' 36"	West
2021	17 May	05:55 22° 00' 36"	East
2021	4 Jul	19:46 21° 33' 00"	West
2021	14 Sep	04:26 26° 45' 36"	East
2021	25 Oct	05:31 18° 23' 24"	West
2022	7 Jan	11:05 19° 13' 12"	East
2022	16 Feb	21:08 26° 16' 12"	West
2022	29 Apr	08:10 20° 36' 00"	East
2022	16 Jun	14:57 23° 11' 24"	West
2022	27 Aug	16:16 27° 19' 12"	East
2022	8 Oct	21:15 17° 58' 48"	West
2022	21 Dec	15:33 20° 07' 48"	East
2023	30 Jan	05:55 24° 57' 36"	West
2023	11 Apr	22:12 19° 28' 48"	East
2023	29 May	05:35 24° 53' 24"	West
2023	10 Aug	01:48 27° 24' 00"	East
2023	22 Sep	13:17 17° 51' 36"	West

ELONGATIONS OF VENUS

Date		Elongation	Type
2015	6 Jun	18:29 45° 23' 40"	East
2015	26 Oct	07:10 46° 26' 29"	West
2017	12 Jan	13:18 47° 08' 46"	East
2017	3 Jun	12:29 45° 51' 59"	West
2018	17 Aug	17:30 45° 55' 40"	East
2019	6 Jan	04:53 46° 57' 22"	West
2020	24 Mar	22:13 46° 04' 39"	East
2020	13 Aug	00:13 45° 47' 28"	West

The Winter skies are particularly striking in the Northern Hemisphere because they are long, become really dark in mid-latitudes and are graced by two of the most distinctive patterns in the night sky – the Plough or Big Dipper, and Orion, the Hunter. From mid-latitudes, the seven stars that make up the pattern of the Plough, are circumpolar, meaning that they never set.

The constellation Ursa Major is invaluable as a guide to other constellations. In the southern part of the sky we see Orion the Hunter, which is also of immense value in identifying other constellations. Unlike Ursa Major, Orion is not circumpolar but when above the horizon is quite unmistakeable.

Come first to Ursa Major with its seven main stars. Two of these, Merak and Dubhe, show the way to Polaris in Ursa Minor, the Little Bear, which is of the second magnitude and very close to the North Celestial Pole. This means that it's of middle brightness and hardly seems to move at all.

Orion, the Hunter

The stars of Orion are actually brighter than those of Ursa Major. The two leaders, Betelgeux and Rigel are not alike. Rigel is brilliant white and at least 40,000 times as luminous as the Sun, whereas Betelgeux is a huge red supergiant. It is not as luminous as Rigel, but at least 15,000 times as powerful as the Sun and is of immense size. Like so many red supergiants it is decidedly variable, and at its best it will become nearly as brilliant as Rigel, though for most of the time it is half a magnitude fainter. Orion has three bright stars making up the Hunter's Belt. Extending southwards from the

1

[1] The night sky in winter, centred on the constellation of Orion, under a brilliant Moon.

[2] The Pleiades cluster of stars lies at a distance of about 440 light years from us in the direction of Taurus, the Bull, which makes it the closest star cluster to Earth. The cluster spans about 43 light years. (ESO/S. Brunier.)

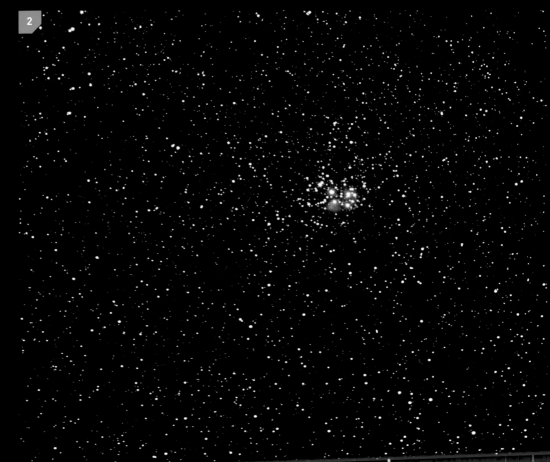

belt is the misty Sword, and the Great Nebula in Orion, a stellar nursery where fresh stars are being formed. It is easily resolved with the naked eye and binoculars show it well. It is a favourite target for astronomical photographers.

In the other direction, upwards (in the northern hemisphere), the Belt stars point to the orange Aldebaran in Taurus the Bull, and extending from Aldebaran is a v-shaped cluster of stars known as the Hyades. Actually, Aldebaran is not a true member of the Hyades, it simply lies about midway between the cluster and us. Beyond Aldebaran we come to the Pleiades, the Seven Sisters, the most famous of all star clusters. With the naked eye it is possible to make out at least seven individual stars and really keen sighted people can see more. It is said that the record is 19!

Almost overhead on winter evenings lies the brilliant Capella in Auriga, the Charioteer. It is slightly yellow in colour and is actually four stars in two binary pairs; the two components are much too close together to be separated with the naked eye or even with a moderate sized telescope.

Close to Capella is a triangle of much fainter stars known as the Heidi, or Kids. At the apex of the triangle is the remarkable Epsilon Auriga, normally of the third magnitude, but every 27.1 years it is eclipsed by a much fainter companion and drops to the fourth magnitude and remains faint for over a year until it recovers.

In a southward (downward) direction, the three stars of Orion's Belt point to Sirius, the brightest star in the sky. It is a pure white star, and when near the horizon it appears to flash various colours because of atmospheric effects.

Orion can also help us to find Procyon in the Little Dog, and the Twins, Castor and Pollux. Leo the Lion is rising in the East and will be better seen in Spring, while the Square of Pegasus is descending in the West and will dominate the sky next Autumn.

The two brightest stars currently visible in the northern hemisphere of the sky are Capella in Auriga and Vega in Lyra, the Lyre. Both are circumpolar from the UK or the northern United States, though when at their lowest they skirt the horizon.

Meteor Showers

There are several meteor showers visible in the winter months from the Northern Hemisphere and, if the Moon and weather work in your favour, these can put on an amazing display against the dark skies of winter. Of these the most reliable is the Geminid meteor shower, radiating from a point near the twins. The shower begins on 7 December and lasts until the 19th with peak activity around the 13th/14th. They are particularly rich: as many as 100 naked-eye meteors can be seen every hour, though strong moonlight will deplete their numbers. We also have the Ursids, radiating from the Little Bear (Ursa Minor). This shower begins by 17 December and ends on Christmas night, it is not generally rich, but is well worth looking out for as it rises to a peak on 22/23 December.

Bright comets are not easy to predict, and at the time of writing, none are expected for a while yet. Of course a really brilliant comet may appear at any time.

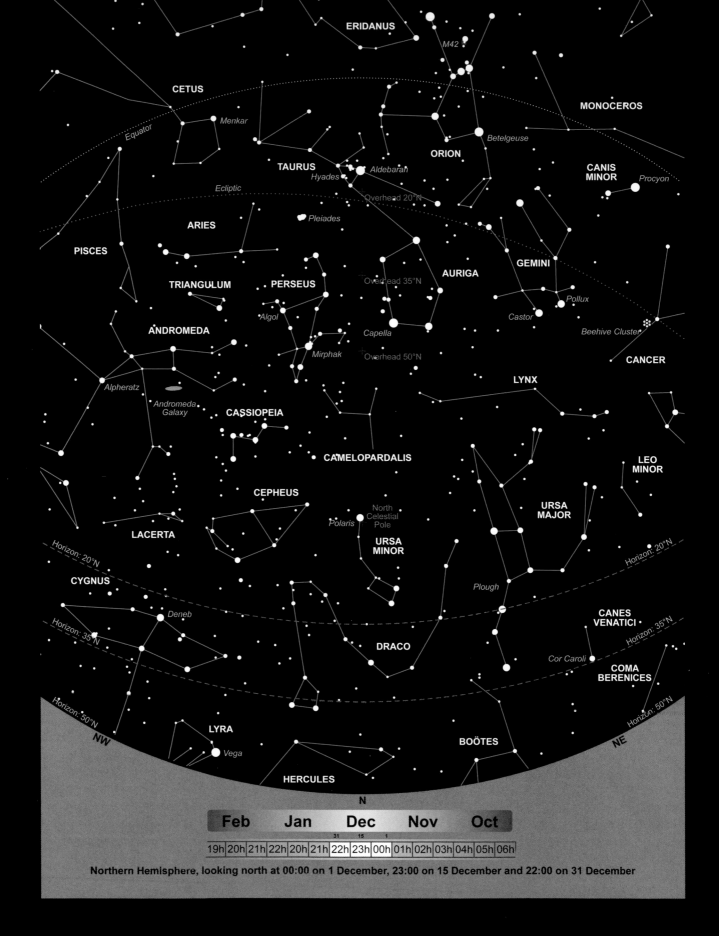

Northern Hemisphere, looking north at 00:00 on 1 December, 23:00 on 15 December and 22:00 on 31 December

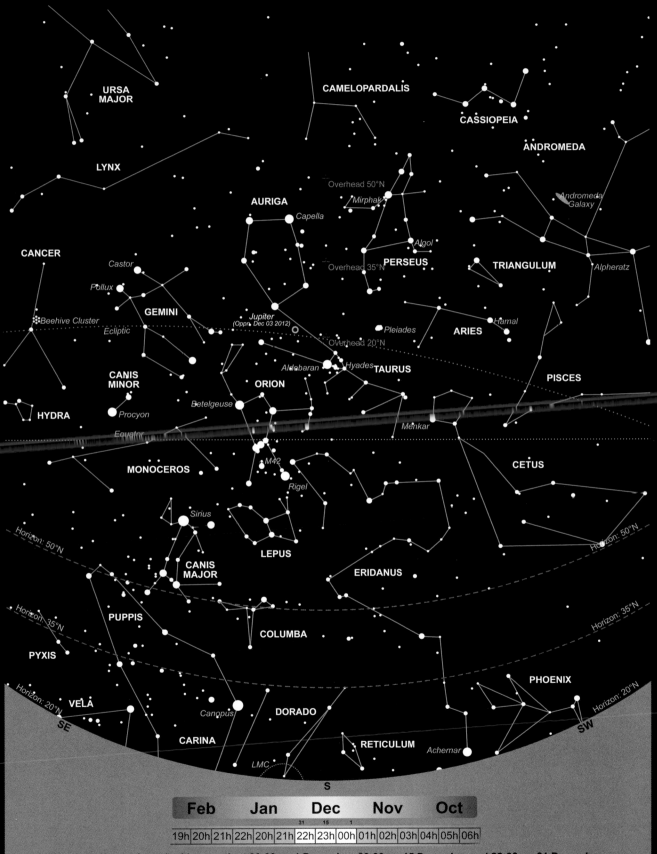

URSA MAJOR

CAMELOPARDALIS

CASSIOPEIA

ANDROMEDA

LYNX

AURIGA

Overhead 50°N

Mirphak

Andromeda
Galaxy

Capella

CANCER

Castor

GEMINI

Pollux

Beehive Cluster

Ecliptic

CANIS
MINOR

HYDRA

Procyon

Betelgeuse

ORION

Equator

MONOCEROS

M42

Rigel

Sirius

CANIS
MAJOR

LEPUS

PUPPIS

PYXIS

Horizon: 50°N

Horizon: 35°N

Horizon: 20°N

VELA

SE

CARINA

Canopus

COLUMBA

DORADO

LMC

Jupiter
(Oppn Dec 03 2012)

Algol

PERSEUS

Overhead 35°N

TRIANGULUM

Alpheratz

Hamal

ARIES

Pleiades

Overhead 20°N

Aldebaran

Hyades

TAURUS

PISCES

Menkar

CETUS

ERIDANUS

Horizon: 50°N

Horizon: 35°N

PHOENIX

Horizon: 20°N

SW

RETICULUM

Achernar

S

Feb	Jan	Dec	Nov	Oct

31 15 1

| 19h | 20h | 21h | 22h | 20h | 21h | 22h | 23h | 00h | 01h | 02h | 03h | 04h | 05h | 06h |

Northern Hemisphere, looking south at 00:00 on 1 December, 23:00 on 15 December and 22:00 on 31 December

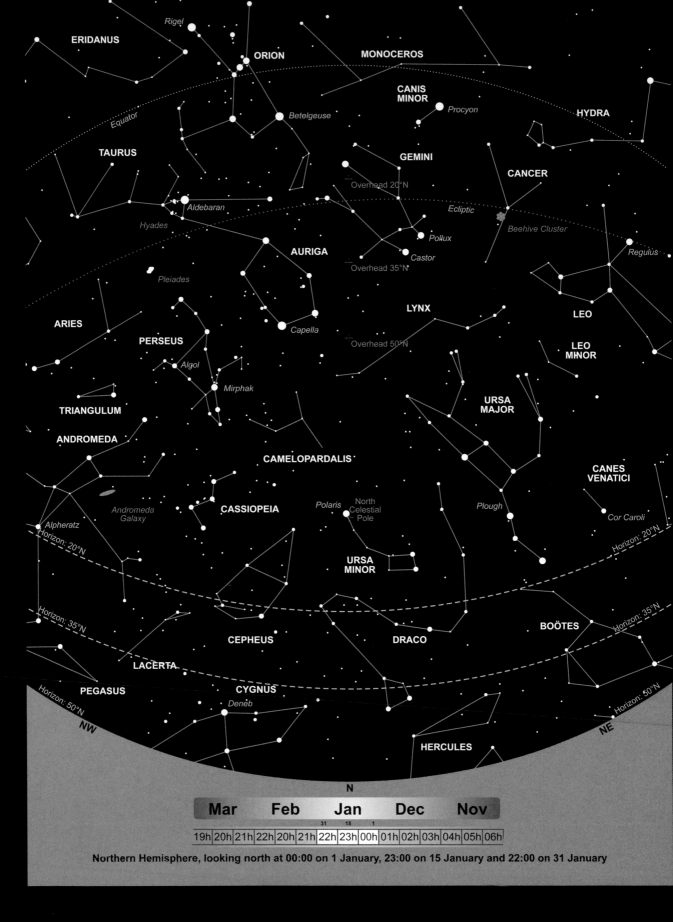

ERIDANUS

Rigel

ORION

MONOCEROS

CANIS
MINOR

Procyon

HYDRA

Equator

Betelgeuse

GEMINI

CANCER

TAURUS

Aldebaran

Overhead 20°N

Ecliptic

Hyades

Pollux

Beehive Cluster

AURIGA

Castor

Overhead 35°N

Regulus

Pleiades

LYNX

LEO

ARIES

Overhead 50°N

Capella

PERSEUS

LEO
MINOR

Algol

Mirphak

URSA
MAJOR

CANES
VENATICI

TRIANGULUM

ANDROMEDA

CAMELOPARDALIS

Cor Caroli

CASSIOPEIA

Polaris

North
Celestial
Pole

Plough

*Andromeda
Galaxy*

Horizon: 20°N

Alpheratz

URSA
MINOR

Horizon: 20°N

Horizon: 35°N

BOÖTES

Horizon: 35°N

CEPHEUS

DRACO

LACERTA

PEGASUS

CYGNUS

Horizon: 50°N

Horizon: 50°N

Deneb

NW

NE

HERCULES

N

Mar	Feb	Jan	Dec	Nov

31 15 1

19h	20h	21h	22h	20h	21h	22h	23h	00h	01h	02h	03h	04h	05h	06h

Northern Hemisphere, looking north at 00:00 on 1 January, 23:00 on 15 January and 22:00 on 31 January

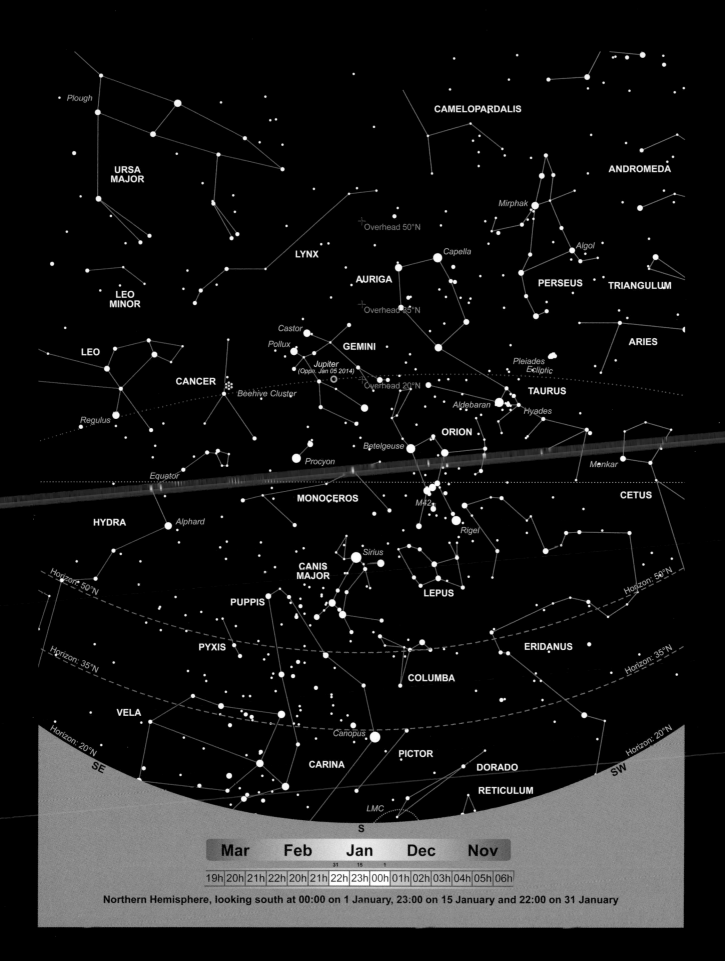

Northern Hemisphere, looking south at 00:00 on 1 January, 23:00 on 15 January and 22:00 on 31 January

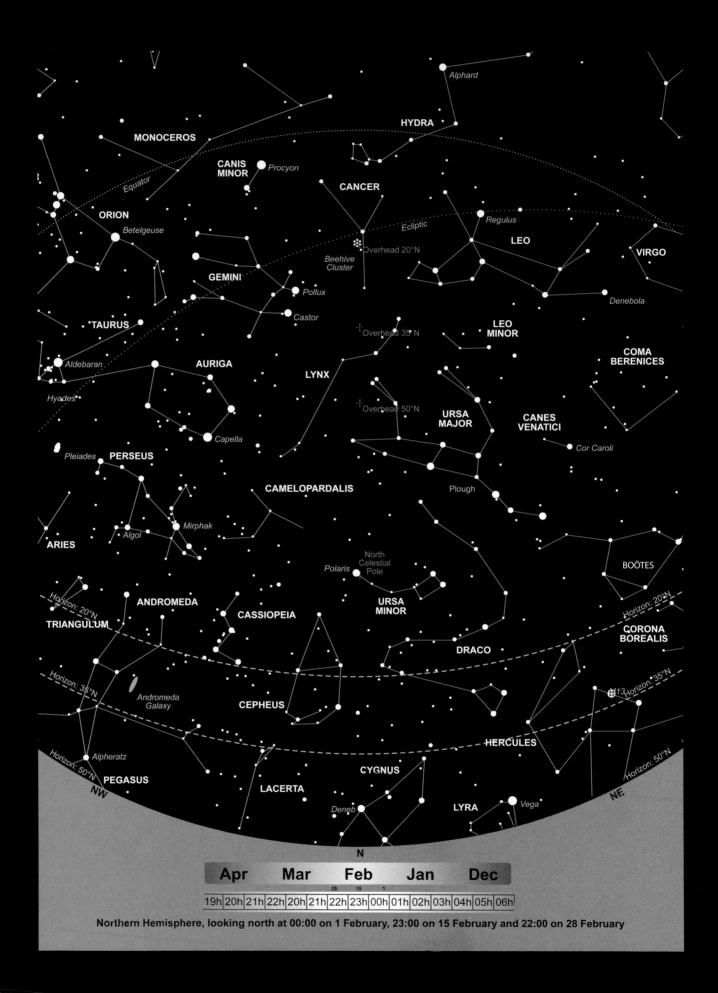

Alphard

HYDRA

MONOCEROS

CANIS
MINOR   *Procyon*

CANCER

Equator

ORION

Betelgeuse

Ecliptic   *Regulus*

Overhead 20°N

LEO

VIRGO

*Beehive
Cluster*

GEMINI

Pollux   Denebola

Castor

TAURUS   Overhead 35°N

LEO
MINOR

Aldebaran   AURIGA

LYNX   COMA
BERENICES

Hyades   Overhead 50°N

URSA
MAJOR   CANES
VENATICI

Capella   Cor Caroli

Pleiades   PERSEUS

Plough

CAMELOPARDALIS

Mirphak

ARIES   *Algol*

North
Celestial
Pole   BOÖTES

Polaris

URSA
MINOR

ANDROMEDA

CASSIOPEIA   Horizon: 20°N

TRIANGULUM   CORONA
BOREALIS

Horizon: 20°N

DRACO   Horizon: 35°N

Horizon: 35°N   *Andromeda
Galaxy*   CEPHEUS

HERCULES

Horizon: 50°N   *Alpheratz*

CYGNUS   Horizon: 50°N

NW   PEGASUS   LACERTA

NE

Deneb   LYRA   *Vega*

N

Apr	Mar	Feb	Jan	Dec

19h	20h	21h	22h	20h	21h	22h	23h	00h	01h	02h	03h	04h	05h	06h

Northern Hemisphere, looking north at 00:00 on 1 February, 23:00 on 15 February and 22:00 on 28 February

Northern Hemisphere, looking south at 00:00 on 1 February, 23:00 on 15 February and 22:00 on 28 February

SPRING (NORTHERN HEMISPHERE)

1 At the end of winter we begin to lose Orion, though some members of the Hunter's retinue are visible well after darkness. Ursa Major is now almost directly overhead. Follow the line of the Great Bear's tail around and you will come to a brilliant orange star, Arcturus in Bootes, the Herdsman, which is slightly above zero magnitude and therefore fractionally brighter than Capella or Vega.

Arcturus sits at the bottom of a huge kite shaped pattern of stars that represents the herdsman. Adjoining Bootes is Corona Borealis – the Northern Crown. Easily identifiable because of the little semicircle of stars of which the brightest is the second magnitude Alphekka. There is one particularly interesting object in Corona Borealis, the variable star R-Coronae Borealis. Generally it is on the fringe of naked-eye visibility, but at unpredictable intervals it fades and becomes so faint that a telescope of some size is needed to show it. This is because

it accumulates clouds of soot in its atmosphere, dims, and remains red-pink until the soot disperses.

Continue the line from the Great Bear's tail to Arcturus and you come to Spica in Virgo, the Virgin, the second largest constellation in the entire sky. The main stars of Virgo are a "Y" shape. Southwest of Virgo, lower in the sky you will be able to find the four stars making up the main quadrilateral of Corvus the Crow. They are only between magnitudes two and three but their pattern is distinctive.

2

[1] Regulus and Mars.

[2] M13 taken with a Canon 10D, 10 x 90 s, ISO 800, Median Stacked Prime Focus Vixen FL-102s, f9 refractor.

[3] The constellation Leo in the Spring skies.

Leo the Lion

High in the south we find the main spring constellation, Leo the Lion. Its main stars make up a reverse question-mark pattern known as the Sickle. The brightest star in the Sickle is Regulus of first magnitude, while Gamma Leonis, to the north, is a lovely double star. East of the Sickle we have the other main part of Leo consisting of a triangle of stars of which the brightest is Denebola. Denebola is now of the second magnitude. It has been claimed that it used to be as bright as Regulus and has now faded, but the evidence for this is very inconclusive.

Castor and Pollux, the Twins

The Twins, Castor and Pollux are still visible. Between them and Regulus lies the dim zodiacal constellation of Cancer, the Crab. It contains no bright stars, but there are two lovely open clusters of which one, Praesepe, the Beehive, is visible with the naked eye and is a fine sight in binoculars.

In a southward (downward) direction, Castor and Pollux point to the reddish, second-magnitude star Alphard, in Hydra the Watersnake. The star is nicknamed the "Solitary One", for there are no other bright stars anywhere near it. Hydra is the largest constellation in the entire sky; it sprawls along eastward from a point close to Orion, but it is lacking in interesting objects.

Lynx is now almost overhead. It adjoins Ursa Major and contains one prominent reddish star, but very little else. It has been said that you need eyes like those of a Lynx to see anything at all. Vega is now rather low in the east, but gaining altitude so that Capella on the opposite side of the Pole Star is dropping down towards the horizon.

Vega will be at its best in Summer and is one member of the so-called "Summer Triangle". Adjoining Vega is Cygnus the Swan, headed by the first magnitude Deneb. The Milky Way is rich round here but will be better seen during the summer. Cassiopeia remains visible, as is Perseus with its famous double cluster. Rising in the east we have Hercules, a large but not bright constellation. Here the main object of interest is the globular cluster M13, the finest globular visible in the northern hemisphere and surpassed only by the southern hemisphere's Omega Centauri and 47 Tucani.

M13 is easy to locate, and is just visible with the naked eye. Binoculars show it well. A small telescope will resolve its outer regions into individual stars.

The Lyrid Meteor Shower

The main recurring meteor shower is that of the Lyrids, lasting from 18 to 25 April. They are usually moderate, but occasionally brilliant as in 1922 and 1982, so it's worthwhile keeping watch on them.

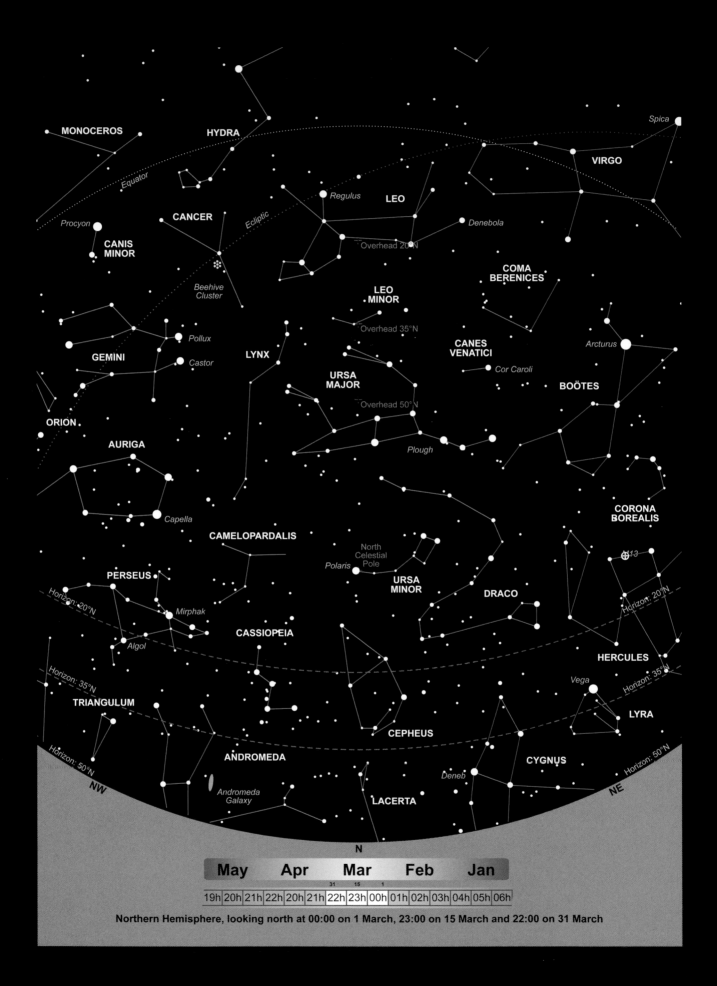

MONOCEROS

HYDRA

Spica

VIRGO

Equator

CANCER

Ecliptic

Regulus

LEO

Denebola

COMA
BERENICES

Procyon

CANIS
MINOR

Overhead 20°N

*Beehive
Cluster*

LEO
MINOR

CANES
VENATICI

Arcturus

Pollux

Overhead 35°N

Cor Caroli

GEMINI

Castor

LYNX

URSA
MAJOR

BOÖTES

ORION

Overhead 50° N

AURIGA

Plough

CORONA
BOREALIS

Capella

CAMELOPARDALIS

North
Celestial
Pole

URSA
MINOR

M13

PERSEUS

Polaris

DRACO

Horizon: 20°N

Horizon: 20°N

Mirphak

CASSIOPEIA

HERCULES

Algol

Vega

Horizon: 35°N

Horizon: 35°N

TRIANGULUM

LYRA

Horizon: 50°N

CEPHEUS

ANDROMEDA

CYGNUS

Horizon: 50°N

NW

Deneb

NE

*Andromeda
Galaxy*

LACERTA

N

May	Apr	Mar	Feb	Jan

31   15   1

| 19h | 20h | 21h | 22h | 20h | 21h | 22h | 23h | 00h | 01h | 02h | 03h | 04h | 05h | 06h |

Northern Hemisphere, looking north at 00:00 on 1 March, 23:00 on 15 March and 22:00 on 31 March

DRACO

BOÖTES

CANES
VENATICI

Cor Caroli

Plough

URSA
MAJOR

Overhead 50°N

LYNX

AURIGA

Capella

Overhead 35°N

LEO
MINOR

CANCER

Castor

Pollux

GEMINI

COMA
BERENICES

LEO

Overhead 20°N

Beehive•Cluster

Ecliptic

Arcturus

Denebola

Regulus

CANIS
MINOR

Procyon

VIRGO

Jupiter

(Oppn. Mar 08 2016)

Equator

Spica

CRATER

HYDRA

Alphard

MONOCEROS

CORVUS

Horizon: 50°N

Sirius

CANIS
MAJOR

Horizon: 50°N

LIBRA

Horizon: 35°N

PUPPIS

PYXIS

Horizon: 35°N

CENTAURUS

VELA

COLUMBA

Horizon: 20°N

Omega
Centauri

Horizon: 20°N

SW

SE

LUPUS

CRUX

Mimosa

Eta Carinae
Nebula

CARINA

Acrux

S

May	Apr	Mar	Feb	Jan

31 15 1

| 19h | 20h | 21h | 22h | 20h | 21h | 22h | 23h | 00h | 01h | 02h | 03h | 04h | 05h | 06h |

Northern Hemisphere, looking south at 00:00 on 1 March, 23:00 on 15 March and 22:00 on 31 March

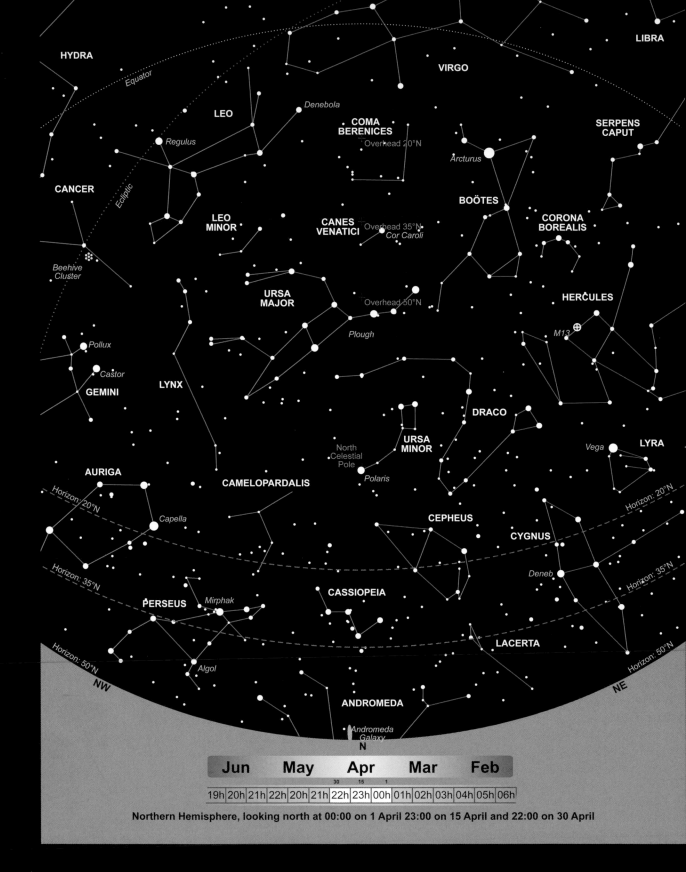

HYDRA

Equator

LIBRA

VIRGO

LEO

Denebola

COMA
BERENICES

Overhead 20°N

SERPENS
CAPUT

Regulus

Arcturus

CANCER

Ecliptic

LEO
MINOR

CANES
VENATICI

Overhead 35°N
Cor Caroli

BOÖTES

CORONA
BOREALIS

Beehive
Cluster

URSA
MAJOR

Overhead 50°N

HERCULES

M13

Pollux

Plough

Castor

LYNX

GEMINI

URSA
MINOR

DRACO

LYRA

Vega

AURIGA

North
Celestial
Pole

Polaris

CAMELOPARDALIS

CEPHEUS

Horizon: 20°N

CYGNUS

Horizon: 20°N

Capella

Deneb

Horizon: 35°N

Horizon: 35°N

PERSEUS

Mirphak

CASSIOPEIA

Horizon: 50°N

LACERTA

Horizon: 50°N

Algol

NW

NE

ANDROMEDA

Andromeda
Galaxy

N

Jun	May	Apr	Mar	Feb

19h	20h	21h	22h	20h	21h	22h	23h	00h	01h	02h	03h	04h	05h	06h

Northern Hemisphere, looking north at 00:00 on 1 April 23:00 on 15 April and 22:00 on 30 April

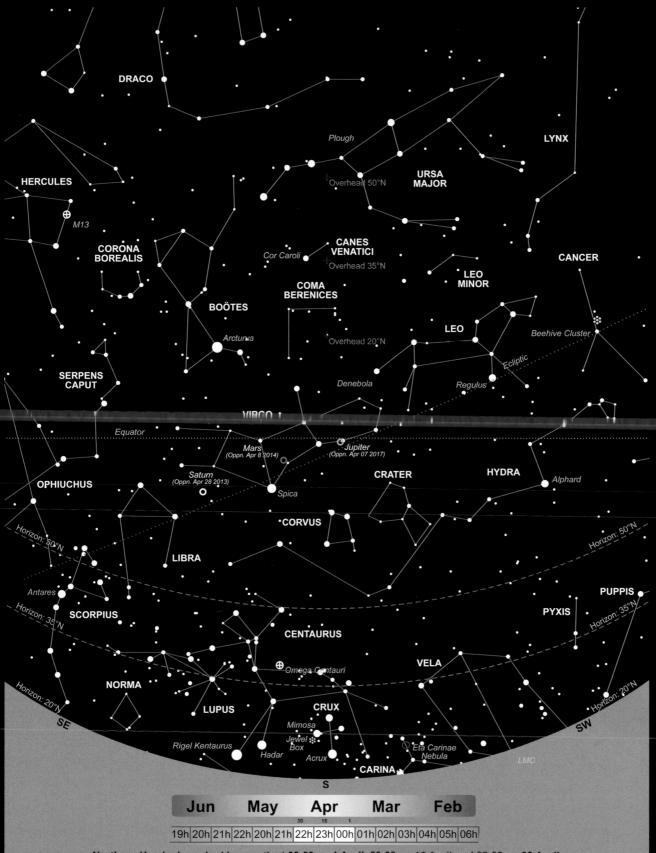

DRACO

LYNX

Plough

Overhead 50°N

URSA
MAJOR

HERCULES

⊕ M13

CORONA
BOREALIS

CANES
VENATICI

Cor Caroli

Overhead 35°N

CANCER

LEO
MINOR

COMA
BERENICES

Beehive Cluster

BOÖTES

LEO

Arcturus

Overhead 20°N

Ecliptic

SERPENS
CAPUT

Denebola

Regulus

VIRGO

Equator

CRATER

HYDRA

Mars
(Oppn. Apr 8 2014)

Jupiter
(Oppn. Apr 07 2017)

Alphard

OPHIUCHUS

Saturn
(Oppn. Apr 28 2013)

Spica

Horizon: 50°N

CORVUS

Horizon: 50°N

LIBRA

PUPPIS

Antares

PYXIS

Horizon: 35°N

Horizon: 35°N

SCORPIUS

CENTAURUS

VELA

NORMA

⊕ Omega Centauri

Horizon: 20°N

Horizon: 20°N

LUPUS

CRUX

SE

Mimosa

SW

Jewel
Box

Eta Carinae
Nebula

Rigel Kentaurus

Hadar

Acrux

LMC

Acrux

CARINA

S

Jun	May	Apr	Mar	Feb

30 15 1

19h	20h	21h	22h	20h	21h	22h	23h	00h	01h	02h	03h	04h	05h	06h

Northern Hemisphere, looking south at 00:00 on 1 April, 23:00 on 15 April and 22:00 on 30 April

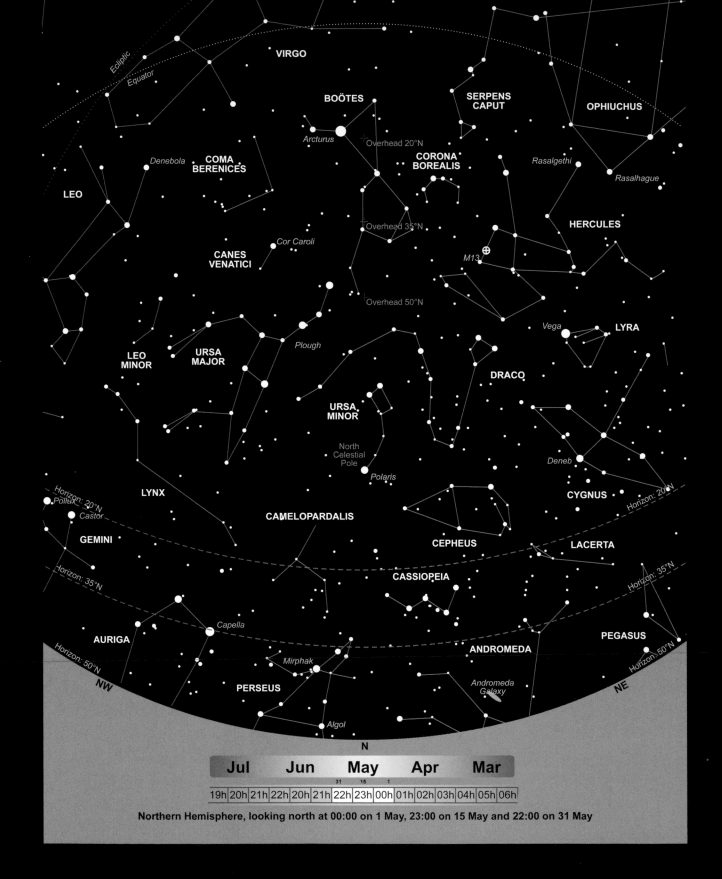

VIRGO

Ecliptic

Equator

BOÖTES

SERPENS
CAPUT

OPHIUCHUS

Arcturus

+ Overhead 20°N

CORONA
BOREALIS

Rasalgethi

Rasalhague

Denebola

COMA
BERENICES

HERCULES

LEO

+ Overhead 35°N

Cor Caroli

CANES
VENATICI

M13 ⊕

Overhead 50°N

Vega

LYRA

LEO
MINOR

URSA
MAJOR

Plough

DRACO

URSA
MINOR

Deneb

North
Celestial
Pole

Polaris

CYGNUS

Horizon: 20°N

LYNX

CAMELOPARDALIS

CEPHEUS

LACERTA

Horizon: 35°N

Pollux Horizon: 20°N

Castor

GEMINI

CASSIOPEIA

Horizon: 35°N

PEGASUS

Horizon: 50°N

Capella

ANDROMEDA

AURIGA

NW

Mirphak

*Andromeda
Galaxy*

NE

PERSEUS

Algol

N

31 15 1

| 19h | 20h | 21h | 22h | 20h | 21h | 22h | 23h | 00h | 01h | 02h | 03h | 04h | 05h | 06h |

Northern Hemisphere, looking north at 00:00 on 1 May, 23:00 on 15 May and 22:00 on 31 May

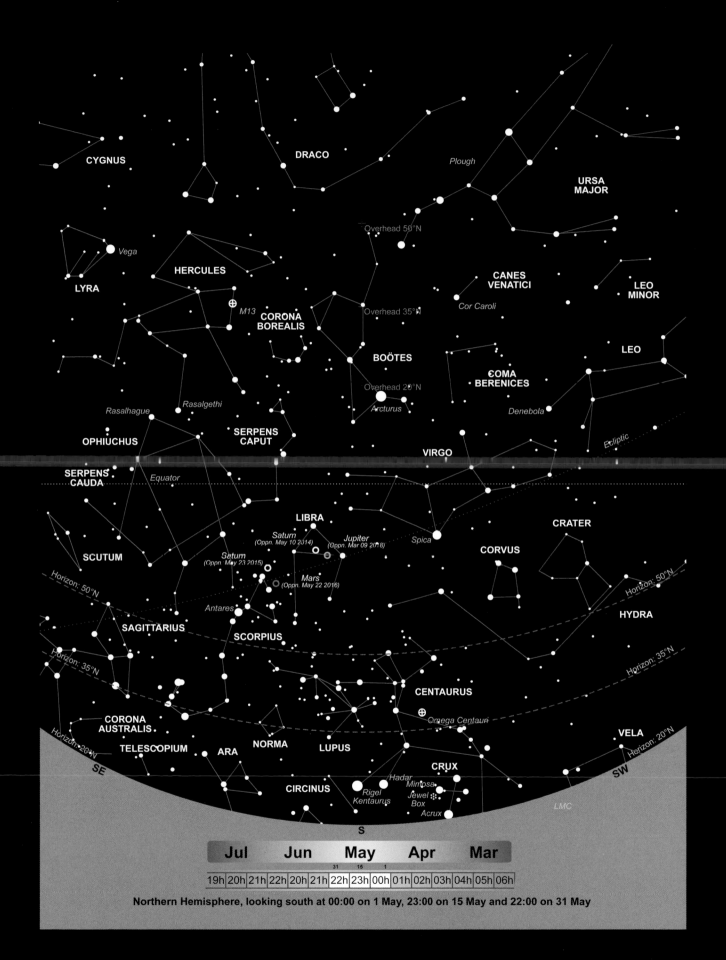

Northern Hemisphere, looking south at 00:00 on 1 May, 23:00 on 15 May and 22:00 on 31 May

By the time that the stars of summer are prominent in the sky, Orion and its retinue have left us. We do have the so-called "Summer Triangle" made up of three bright stars, Vega in Lyra the Lyre or Harp, Deneb in Cygnus the Swan and Altair in Aquila the Eagle. In fact, the stars of the Summer Triangle are unrelated and are in different constellations. In a *Sky at Night* programme many years ago Patrick Moore referred to these stars as the Summer Triangle and for some reason the name caught on and is now used widely.

Vega is now very near the zenith, which means that Capella is low down and grazing the horizon. Deneb in Cygnus is a particularly powerful star, but very distant, and if it were as close as Sirius it would cast strong shadows. Cygnus, often called the Northern Cross, contains many interesting objects and is very rich, because it is crossed by a particularly bright area of the Milky Way. Of the five stars making up the cross of Cygnus, the southernmost and faintest, Albireo, Beta Cygni, is a glorious double star, with a golden yellow primary and azure blue companion. A small telescope will show the pair well and there is little doubt it is the loveliest colour double in the entire sky. Between it and Sadr, the central star of the Cross, there is a well known variable star, Chi Cygni, which can rise to the third magnitude but becomes very faint at minimum. It is a star of the Mira type and has an exceptionally long period.

Altair

The third member of the Summer Triangle, Altair, is identifiable not only by its brightness but because it is flanked by prominent stars on either side of it. Below it is a line of stars, of which the middle member, Eta Aquilae, is the brightest Cepheid variable in the sky apart from Delta Cephei itself. Its period is seven days, whereas Delta Cephei has a period of five days. This shows that of the two, Eta Aquilae is the more luminous.

As we have noted, Ursa Major is high up, and the region between it and the Twins is filled mainly by the large dim Lynx. Hercules, roughly between Vega and Arcturus, is also high up and contains the globular M13. Note also Alpha Herculis, a red supergiant which varies between magnitudes three and four. It has a small companion star, which looks green mainly because of contrast. Close by, we find the large constellation of Ophiuchus, the Serpent Bearer, with one bright star, Rasalhague, intertwined with Ophiuchus in Serpens the Serpent. The two are presumably struggling and the Serpent has come off worse since it has been pulled in half and there are two separate parts of the constellation – the head (Caput) and the tail (Cauda). One star worth looking at in the tail is Theta Serpentis, which is a fine double whose components appear to be virtually identical.

The Scorpion and the Archer

Look now in the southern sky not far from the horizon. Here we have two splendid Zodiacal constellations Scorpius, the Scorpion, and Sagittarius, the Archer. The leading star of Scorpius is Antares, a huge red supergiant which is a lovely sight even in binoculars. Like Alpha Herculis it has a faint companion that looks greenish by contrast with its red primary. Scorpius itself is marked by a line of bright stars of which the leader is Antares. Unfortunately it is never at its best seen from British latitudes because it is low down and the Scorpion's sting is very difficult to see. This whole area is worth sweeping with binoculars because the Milky Way is rich here and there are many clusters.

Ajoining Scorpius is Sagittarius, the Archer, which contains the lovely star clouds that block our views of the centre of the Galaxy. They are never well seen from Britain, but from southern countries they can cast shadows.

The Perseid Meteor Shower

The best, but by no means the only, summer meteor shower is that of the Perseids which begins at the end of July and reaches a maximum around 12/13 August. This is the most reliable shower of the year. If you wait for a dark sky you are almost certain to see a number of bright meteors. They are not difficult to photograph but you do need considerable luck in having the camera pointing in the right place at the right time.

[1] Albireo, viewed with the naked eye appears as a single star, but with a small telescope it can be seen to be a beautiful double star. It is the fifth brightest star in the sky.

[2] The Summer Triangle: Deneb, the brightest star in Cygnus in the top left quarter of the image, makes up the triangle with Vega, in Lyra, at right, and Altair in Aquila down towards the horizon.

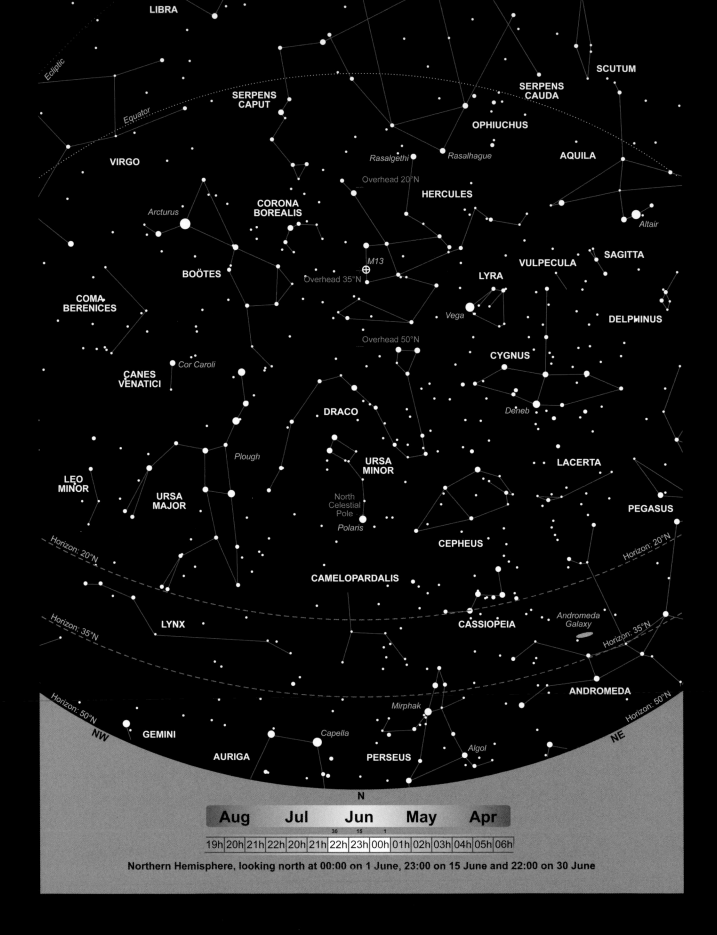

Northern Hemisphere, looking north at 00:00 on 1 June, 23:00 on 15 June and 22:00 on 30 June

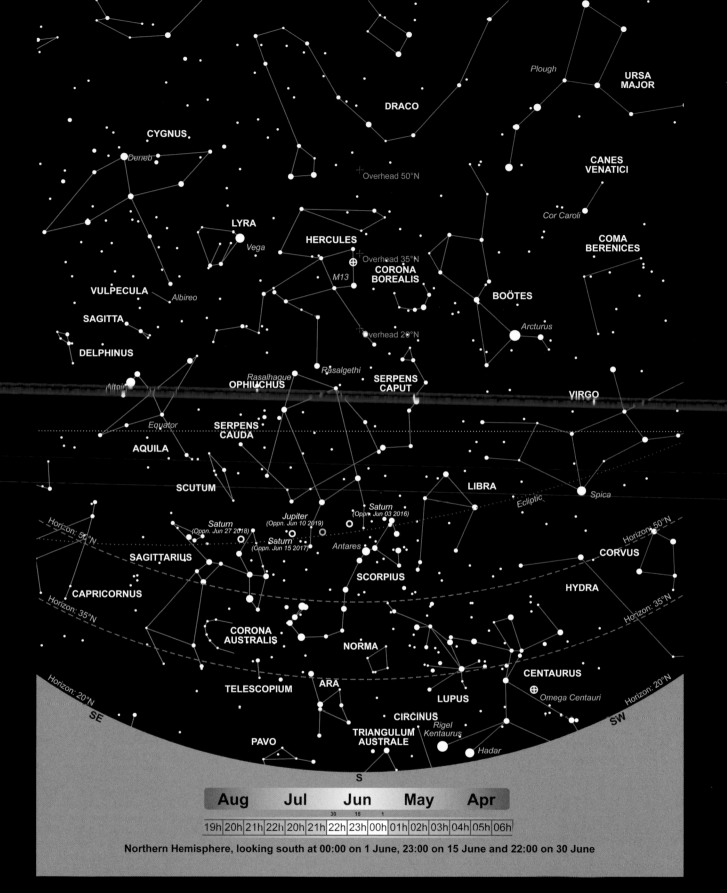

Northern Hemisphere, looking south at 00:00 on 1 June, 23:00 on 15 June and 22:00 on 30 June

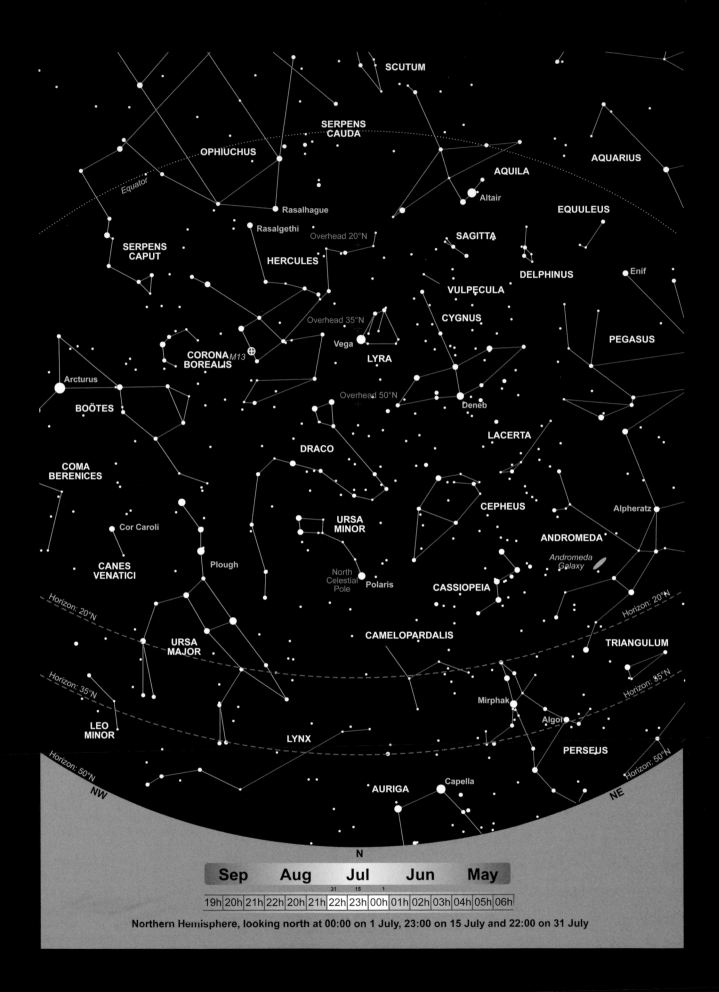

SCUTUM

SERPENS
CAUDA

OPHIUCHUS

AQUARIUS

Equator

AQUILA
Altair

EQUULEUS

Rasalhague

SERPENS
CAPUT

Rasalgethi

Overhead 20°N

SAGITTA

Enif

HERCULES

VULPECULA

DELPHINUS

CYGNUS

PEGASUS

Overhead 35°N

Vega

LYRA

Arcturus

CORONA
BOREALIS *M13*

Overhead 50°N

Deneb

LACERTA

BOÖTES

COMA
BERENICES

DRACO

CEPHEUS

ANDROMEDA

Alpheratz

*Andromeda
Galaxy*

Cor Caroli

URSA
MINOR

CANES
VENATICI

Plough

North
Celestial
Pole

Polaris

CASSIOPEIA

Horizon: 20°N

Horizon: 20°N

TRIANGULUM

CAMELOPARDALIS

URSA
MAJOR

Horizon: 35°N

Horizon: 35°N

LEO
MINOR

Mirphak

LYNX

Algol

Horizon: 50°N

PERSEUS

Horizon: 50°N

NW

AURIGA

Capella

NE

N

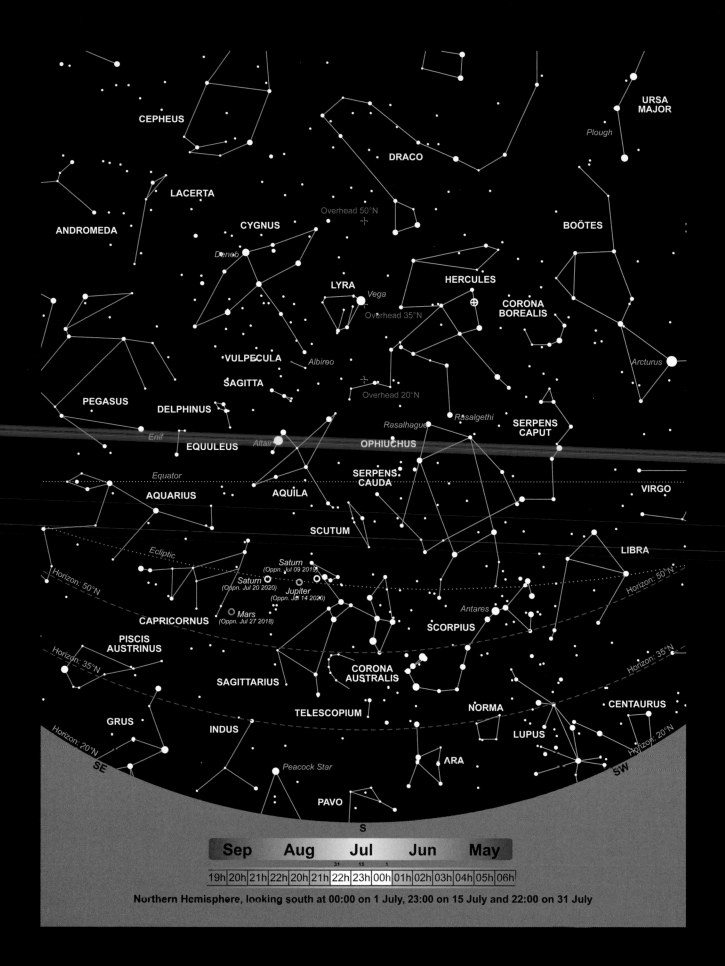

CEPHEUS

URSA MAJOR

Plough

DRACO

LACERTA

ANDROMEDA

CYGNUS

Deneb

BOÖTES

LYRA

Vega

HERCULES

Overhead 50°N

Overhead 35°N

CORONA BOREALIS

VULPECULA

Albireo

SAGITTA

Overhead 20°N

Rasalgethi

Arcturus

PEGASUS

DELPHINUS

Rasalhague

SERPENS CAPUT

Enif

EQUULEUS

Altair

OPHIUCHUS

Equator

SERPENS CAUDA

AQUARIUS

AQUILA

VIRGO

SCUTUM

LIBRA

Ecliptic

Horizon: 50°N

Saturn
(Oppn. Jul 09 2019)

Saturn
(Oppn. Jul 20 2020)

Jupiter
(Oppn. Jul 14 2020)

Horizon: 50°N

CAPRICORNUS

Mars
(Oppn. Jul 27 2018)

Antares

SCORPIUS

PISCIS AUSTRINUS

Horizon: 35°N

CORONA AUSTRALIS

Horizon: 35°N

SAGITTARIUS

CENTAURUS

TELESCOPIUM

NORMA

GRUS

INDUS

LUPUS

Horizon: 20°N

ARA

Horizon: 20°N

SE

SW

Peacock Star

PAVO

S

Sep	Aug	Jul	Jun	May

31 15 1

19h	20h	21h	22h	20h	21h	22h	23h	00h	01h	02h	03h	04h	05h	06h

Northern Hemisphere, looking south at 00:00 on 1 July, 23:00 on 15 July and 22:00 on 31 July

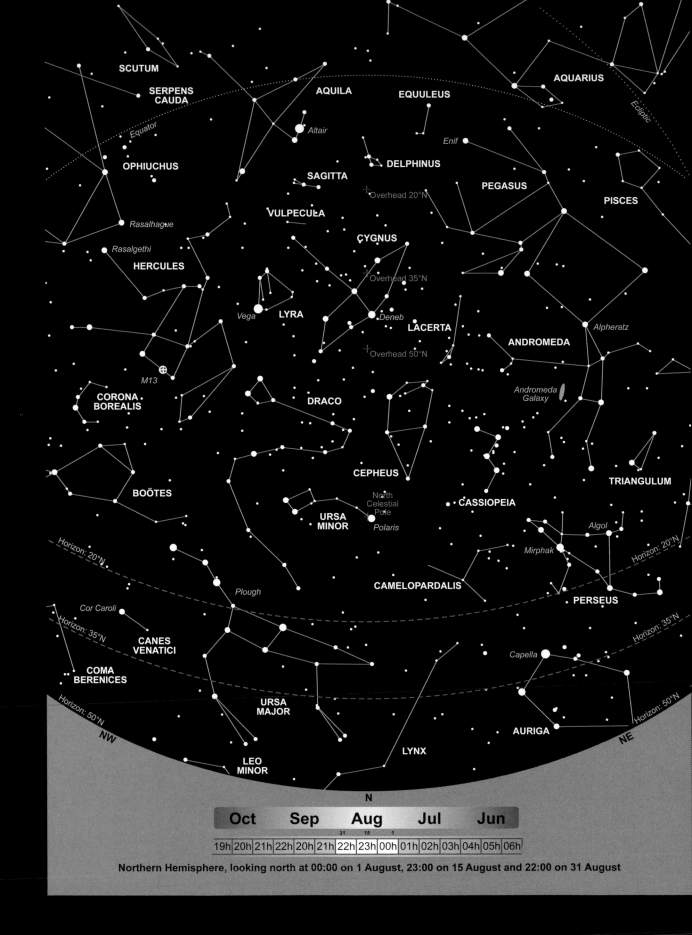

Northern Hemisphere, looking north at 00:00 on 1 August, 23:00 on 15 August and 22:00 on 31 August

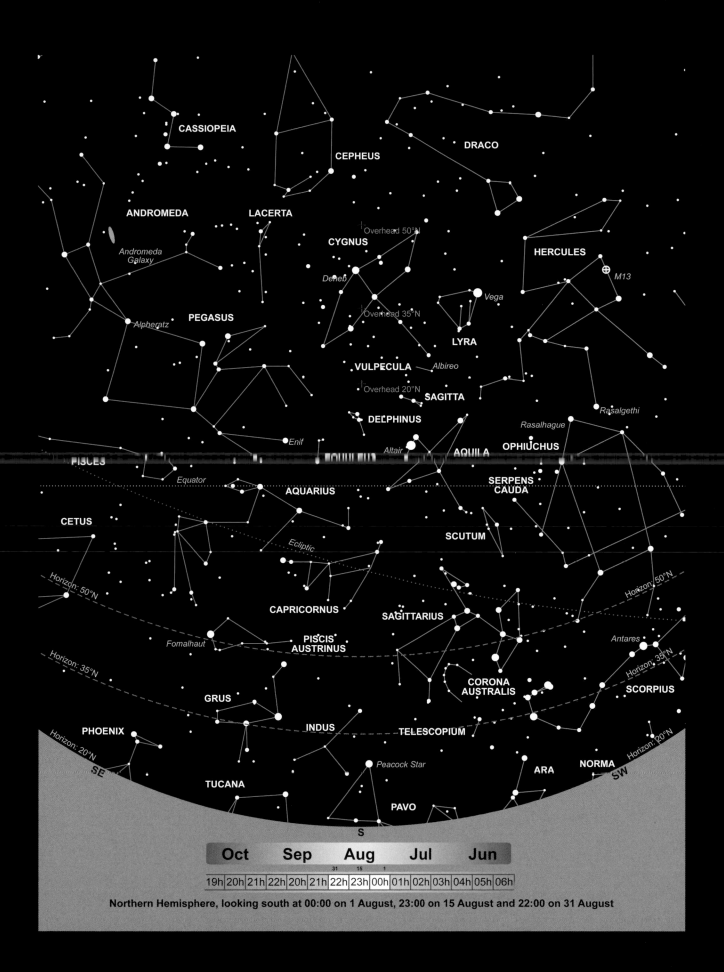

CASSIOPEIA

CEPHEUS

DRACO

ANDROMEDA

LACERTA

Andromeda Galaxy

CYGNUS

Overhead 50°N

HERCULES

⊕ M13

Deneb

Vega

Overhead 35°N

PEGASUS

Alpheratz

LYRA

VULPECULA

Albireo

Overhead 20°N

SAGITTA

DELPHINUS

Enif

Altair

AQUILA

OPHIUCHUS

Rasalhague

Rasalgethi

PISCES

Equator

SERPENS CAUDA

CETUS

AQUARIUS

Ecliptic

SCUTUM

Horizon: 50°N

Horizon: 50°N

CAPRICORNUS

SAGITTARIUS

Antares

Horizon: 35°N

Fomalhaut

PISCIS AUSTRINUS

Horizon: 35°N

SCORPIUS

GRUS

CORONA AUSTRALIS

PHOENIX

INDUS

TELESCOPIUM

Horizon: 20°N

Horizon: 20°N

SE

Peacock Star

ARA

NORMA

SW

TUCANA

PAVO

S

Oct	Sep	Aug	Jul	Jun

31   15   1

| 19h | 20h | 21h | 22h | 20h | 21h | 22h | 23h | 00h | 01h | 02h | 03h | 04h | 05h | 06h |

Northern Hemisphere, looking south at 00:00 on 1 August, 23:00 on 15 August and 22:00 on 31 August

1

It is often said that from the northern hemisphere the autumn stars are the least interesting of the year. We have seen the best of the Summer Triangle, and Orion has yet to make its entrance, but there is still plenty to see. Vega is still high up, which means that Capella is low down. The main autumn constellations are Pegasus and Andromeda, the constellation which contains the magnificent Andromeda spiral galaxy.

In mythology Pegasus was a flying horse, and in the sky its main stars make up a square which is easy to locate, although maps always make it look smaller and brighter than it really is. For some reason the top left-hand star of the square, Alpheratz, which is obviously part of the Pegasus pattern, was given a free transfer to the adjacent constellation of Andromeda by the International Astronomical Union. What was Delta Pegasi, is now known as Alpha Andromedae.

The stars of the square are not alike. Three of them, Alpheratz, Gamma Pegasi (Algenib) and Alpha Pegasi (Markab), are bluish white. The top right-hand star Beta Pegasi (Scheat) is obviously orange and is a semi-regular variable ranging between magnitudes 2¼ and 2¾. This means it is sometimes brighter than Markab and sometimes fainter with an approximate period of 39 days.

Outside the square to the west is Epsilon Pegasi, Enif, and near here there is the fine globular cluster, M15, which is easily seen with binoculars or a small telescope. Look and see how many stars you can observe inside the square. From a city, you may seen none at all. Anywhere between four and 12 is good, while above 12 means you have excellent skies.

The First Extrasolar Planet

Before leaving Pegasus, between the top right-hand star, the orange Scheat and the lower right-hand star, Markab, not far away from the middle of a line joining these two, is the

2

[1] Mira is a double star. Mira A is the bright red variable giant star at right, and its companion star a small hot white dwarf is at left.

[2] M31, the Andromeda galaxy, photographed by Ian Sharp.

[3] The sky in Autumn, with the Square of Pegasus.

inconspicuous star 51 Pegasi, which was the first star found to be attended by a planet.

This discovery, made only in the 1990s, altered our thinking and we now know of hundreds of extrasolar planets, but most have been discovered indirectly and only in a few cases has the planet actually been seen.

Downwards (south), Scheat and Markab point to Fomalhaut in the Southern Fish, the southernmost of the first magnitude stars visible from Britain. It is 22 light years away and 13 times as luminous as our Sun. We know it is a planetary centre, and one planet, Fomalhaut B, has been imaged.

Mira

A line drawn from Scheat in the upper right of the square, through Algenib in the lower left, extended for two times the distance again, will land you in the constellation of Cetus, the Whale or Sea Monster. The most famous object here is the variable star Mira. It has a period of 332 days and ranges from magnitudes two to nine. At its best it is prominent to the naked eye and is clearly red. On average it remains visible only for a few weeks in the year. Like all stars of its type the maximum magnitude is never really predictable and there are times when the star never rises above magnitude three. Patrick Moore has seen it reach the first magnitude for a few nights. This was the first variable of its type to be identified. We now know that Mira variables are very common indeed. Most are a long way away and very few reach naked-eye visibility.

The Andromeda Galaxy

Come back to Andromeda, and we see a line of stars extending away from Alpheratz, of which two are of special interest. Almaak (Gamma Andromedae) is a lovely double star, while the orange Mirak (Beta Andromedae) is a guide to M31 – the great spiral galaxy in Andromeda, the nearest large external galaxy. It is easily visible with the naked eye when you know where to look, but in a small telescope it tends to be a disappointment, because it lies at an unfavourable angle to us and its outer regions are surprisingly faint. What we see with the naked eye or binoculars is just the bright core of M31. This was the first galaxy shown to be an external system rather than part of the Milky Way. The discovery was made by Edwin Hubble over 80 years ago. He found Cepheid variables using the Mount Wilson 100-inch telescope, inside galaxies, and these useful stars give away their distances by their behaviour. This is probably the most important astronomical discovery of modern time. It showed our Galaxy is only one of many and is not even exceptional; it may be somewhat smaller than M31.

The Andromeda spiral has two satellite galaxies M32 and M110, easily seen with binoculars. The whole group is a favourite target for astronomical photographers.

Autumn Meteors

There are several autumn meteor showers. The Draconids peak on 10 October, are usually weak but do display occasional storms. They are associated with the periodical comet Giacobini Zinner. The Orionids, lasting from 16-27 October are associated with Halley's Comet and are characterised by swift meteors leaving fine trails.

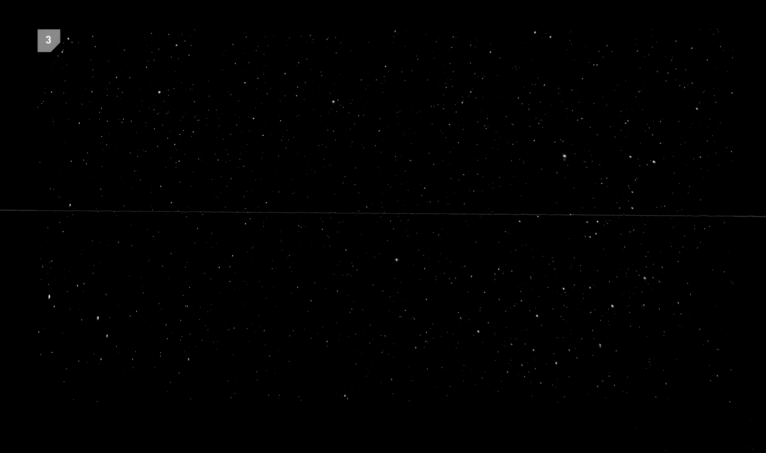

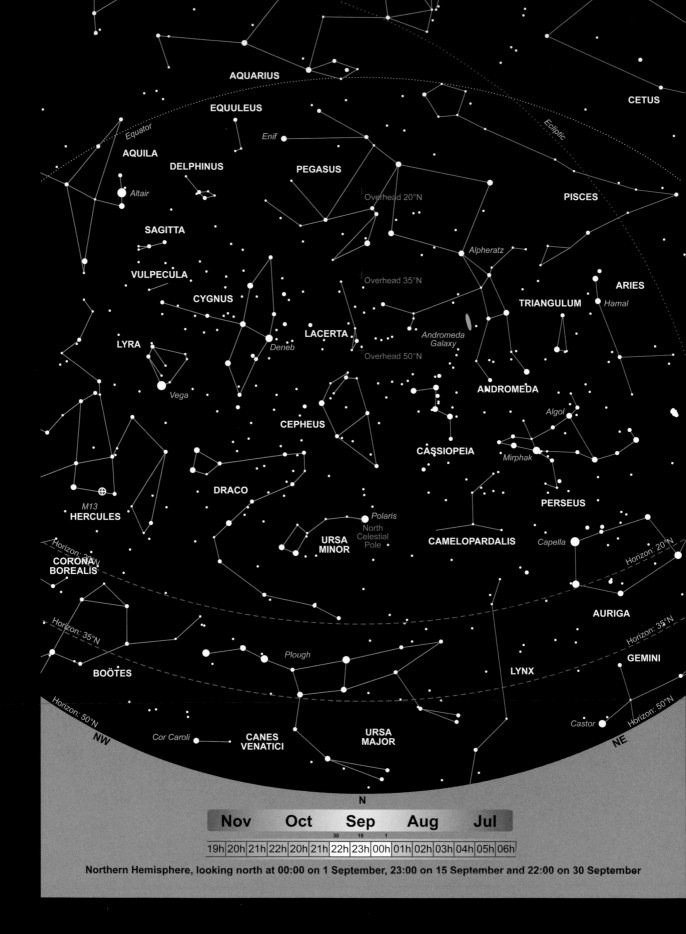

Northern Hemisphere, looking north at 00:00 on 1 September, 23:00 on 15 September and 22:00 on 30 September

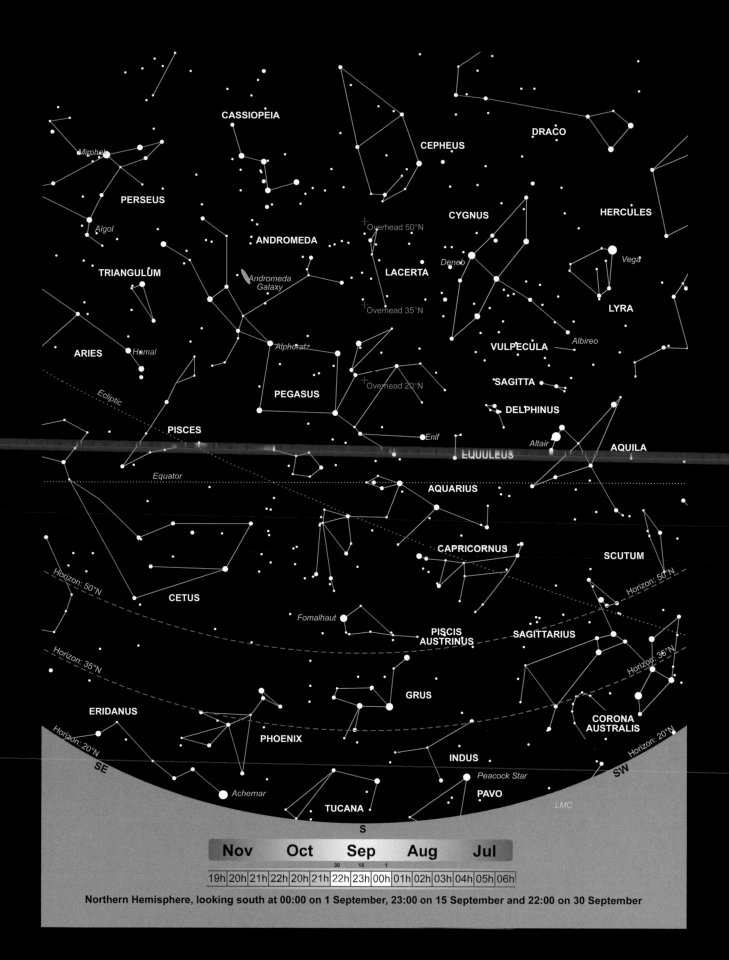

CASSIOPEIA

CEPHEUS

DRACO

PERSEUS

Mirphak

Algol

CYGNUS

HERCULES

ANDROMEDA

Deneb

Vega

TRIANGULUM

LACERTA

Andromeda Galaxy

LYRA

Alpheratz

VULPECULA

Albireo

ARIES

Hamal

Overhead 50°N

Overhead 35°N

SAGITTA

Overhead 20°N

DELPHINUS

PEGASUS

Enif

Altair

AQUILA

PISCES

Ecliptic

EQUULEUS

AQUARIUS

Equator

CAPRICORNUS

SCUTUM

Horizon: 50°N

CETUS

Horizon: 50°N

Fomalhaut

PISCIS AUSTRINUS

SAGITTARIUS

Horizon: 35°N

Horizon: 35°N

GRUS

ERIDANUS

CORONA AUSTRALIS

Horizon: 20°N

PHOENIX

Horizon: 20°N

SE

INDUS

SW

Peacock Star

Achernar

PAVO

TUCANA

LMC

S

Nov	Oct	Sep	Aug	Jul

30 15 1

| 19h | 20h | 21h | 22h | 20h | 21h | 22h | 23h | 00h | 01h | 02h | 03h | 04h | 05h | 06h |

Northern Hemisphere, looking south at 00:00 on 1 September, 23:00 on 15 September and 22:00 on 30 September

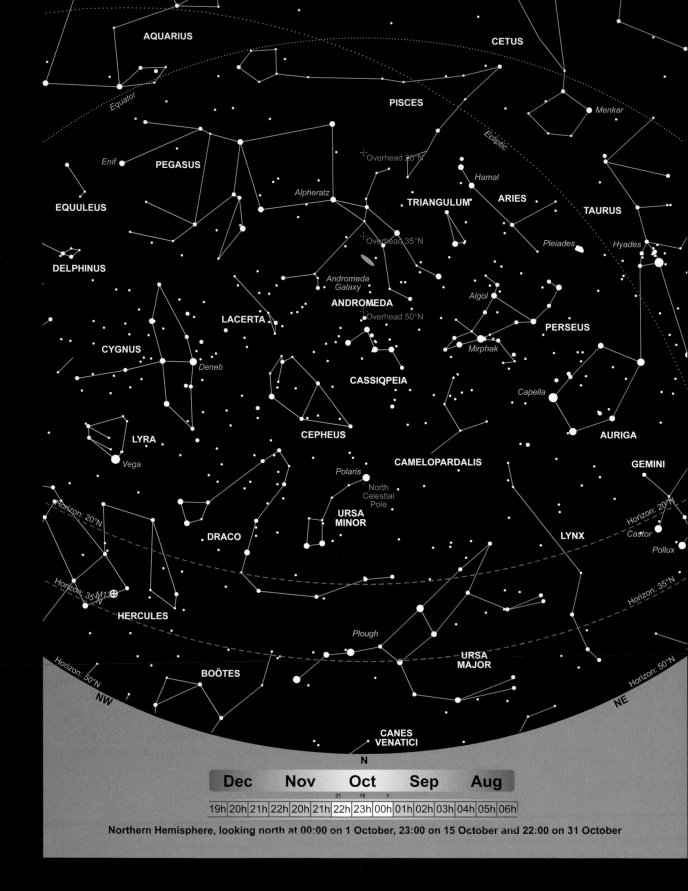

AQUARIUS

CETUS

Equator

PISCES

Menkar

Enif

PEGASUS

+ Overhead 20°N

Hamal

ARIES

TAURUS

Ecliptic

EQUULEUS

Alpheratz

TRIANGULUM

+ Overhead 35°N

Pleiades

Hyades

DELPHINUS

Andromeda Galaxy

ANDROMEDA

Algol

PERSEUS

+ Overhead 50°N

LACERTA

CYGNUS

Deneb

CASSIQPEIA

Mirphak

CEPHEUS

Capella

LYRA

AURIGA

Vega

CAMELOPARDALIS

GEMINI

Polaris

North
Celestial
Pole

Horizon: 20°N

Castor

URSA
MINOR

Pollux

DRACO

LYNX

Horizon: 35°N

Horizon: 20°N

Horizon: 35°N M13

HERCULES

Plough

URSA
MAJOR

Horizon: 50°N

Horizon: 50°N

NW

BOÖTES

NE

CANES
VENATICI

N

Dec	Nov	Oct	Sep	Aug

31 15 1

| 19h | 20h | 21h | 22h | 20h | 21h | 22h | 23h | 00h | 01h | 02h | 03h | 04h | 05h | 06h |

Northern Hemisphere, looking north at 00:00 on 1 October, 23:00 on 15 October and 22:00 on 31 October

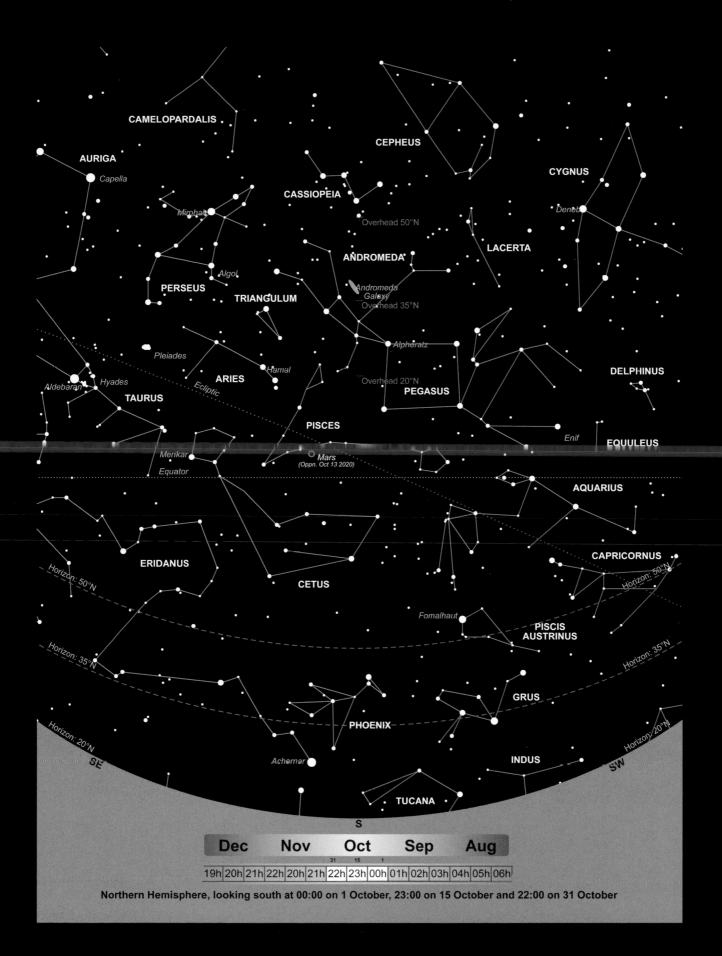

CAMELOPARDALIS

AURIGA

Capella

CEPHEUS

CYGNUS

Deneb

CASSIOPEIA

Mirphak

Overhead 50°N

ANDROMEDA

LACERTA

PERSEUS

Algol

TRIANGULUM

*Andromeda
Galaxy*

Overhead 35°N

Pleiades

Alpheratz

Ecliptic

DELPHINUS

ARIES

Overhead 20°N

Harnal

Hyades

Aldebaran

TAURUS

PEGASUS

Enif

EQUULEUS

PISCES

Menkar

Mars
(Oppn. Oct 13 2020)

Equator

AQUARIUS

CAPRICORNUS

Horizon: 50°N

ERIDANUS

CETUS

Horizon: 50°N

Fomalhaut

PISCIS
AUSTRINUS

Horizon: 35°N

Horizon: 35°N

GRUS

Horizon: 20°N

PHOENIX

INDUS

Horizon: 20°N

SE

Achernar

SW

TUCANA

S

Dec	Nov	Oct	Sep	Aug

31   15   1

| 19h | 20h | 21h | 22h | 20h | 21h | 22h | 23h | 00h | 01h | 02h | 03h | 04h | 05h | 06h |

Northern Hemisphere, looking south at 00:00 on 1 October, 23:00 on 15 October and 22:00 on 31 October

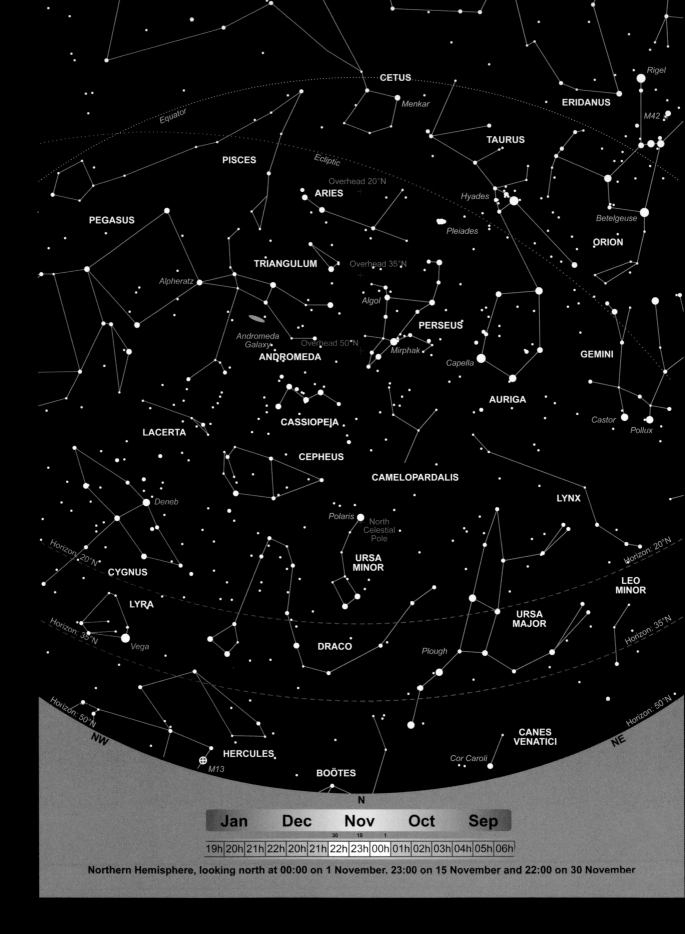

CETUS

Menkar

ERIDANUS

Rigel

M42

TAURUS

Equator

PISCES

Ecliptic

Overhead 20°N

ARIES

Hyades

Betelgeuse

PEGASUS

Pleiades

ORION

TRIANGULUM

Overhead 35°N

Alpheratz

Algol

PERSEUS

GEMINI

Andromeda Galaxy

Overhead 50°N

Mirphak

ANDROMEDA

Capella

AURIGA

CASSIOPEIA

Castor

LACERTA

CEPHEUS

Pollux

CAMELOPARDALIS

LYNX

Deneb

Polaris

North Celestial Pole

Horizon: 20°N

CYGNUS

URSA MINOR

LEO MINOR

LYRA

URSA MAJOR

Horizon: 35°N

Horizon: 35°N

Vega

DRACO

Plough

Horizon: 50°N

Horizon: 50°N

NW

CANES VENATICI

NE

HERCULES

Cor Caroli

⊕ *M13*

BOÖTES

N

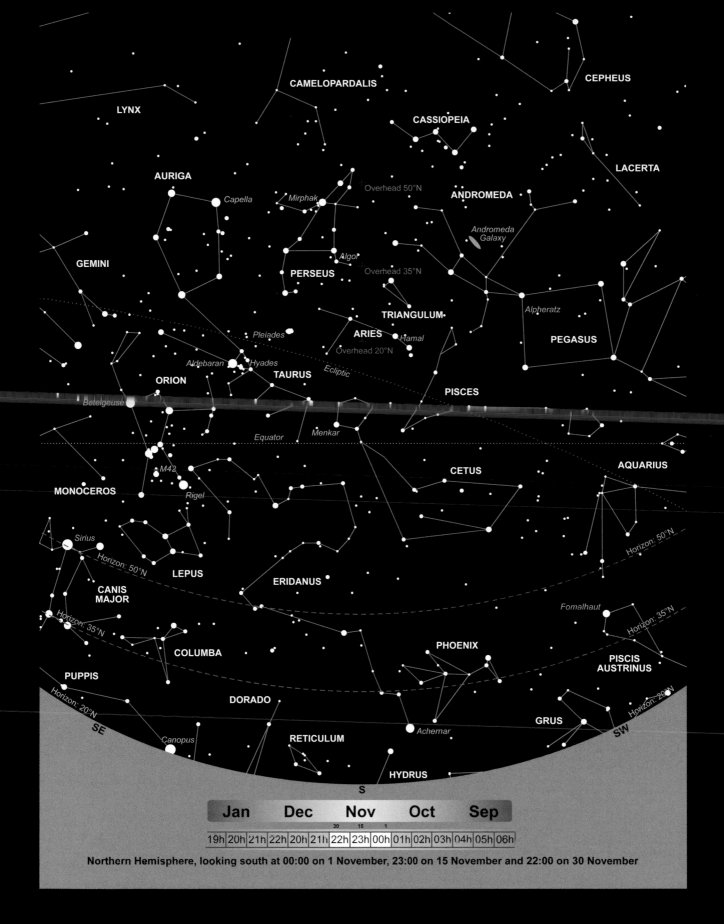

Northern Hemisphere, looking south at 00:00 on 1 November, 23:00 on 15 November and 22:00 on 30 November

WINTER (SOUTHERN HEMISPHERE)

Winter in the southern hemisphere is particularly notable because the star clouds in Sagittarius, the Archer, are almost at the zenith and therefore at their very best. Adjoining the Archer is Scorpius, with the bright red supergiant star, Antares. Note the sting, now practically at the zenith and containing a prominent pair of stars, Shaula (Lambda Scorpii) and Lesath (Upsilon Scorpii).

Shaula is only just below the first magnitude and is not genuinely associated with Lesath: in fact Lesath is much the more remote and luminous of the two. There are two lovely open clusters in Scorpius, M6 and M7, both are fine sights in binoculars and absolutely glorious when viewed with a small telescope using a low power.

Look now toward the north of the summer triangle. Vega is at a reasonable altitude, but Deneb is right on the horizon. Altair is higher up. Trace down the line of stars marking the constellation of Aquila and you will come to a very rich region of the Milky Way together with one lovely star cluster, M11, known as the Wild Duck Cluster.

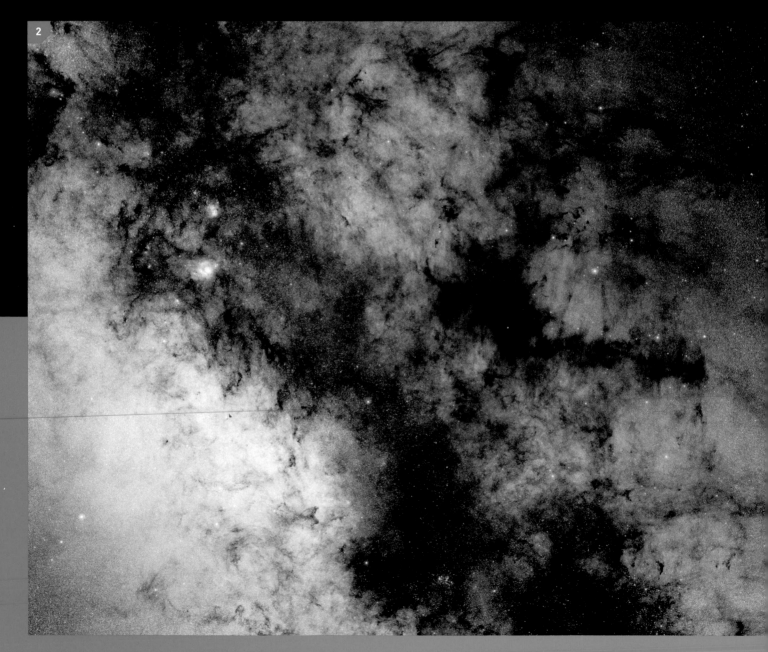

Delphinus, the Dolphin

Below and to the east of Altair look for Delphinus, the Dolphin, which is sometimes confused with the Pleiades open cluster in Taurus, despite not being as bright or as compact as the cluster. Next, let us find the Southern Birds. There are several of these – Grus the Crane, Pavo the Peacock, and Tucana the Tucan. The most distinctive of these is Grus, where there are two second magnitude stars Alnair and Beta Gruis, and a whole line of fainter stars which give the impression of being wide doubles, though they are merely optical pairs.

Pavo contains a star called Kappa Pavonis which, for a long time, was regarded as a Cepheid variable of a type once known as a Type II Cepheid. This is much less luminous than a classic Cepheid such as Delta Cephei or Eta Aquilae.

It was this which initially misled Edwin Hubble when he was trying to measure the distance of the Andromeda Galaxy. He used Cepheids as "standard candles" but he did not know and had no means of knowing that there were two separate kinds of Cepheid, and he had picked the wrong one! This is why his initial estimates for the distances of the spiral galaxies were much too low, though still great enough to demonstrate that they were remote systems and not inside our own galaxy.

Tucana, the Toucan

Tucana has one splendid globular cluster 47 Tucanae, which appears to be silhouetted against the Small Magellanic Cloud. This is highly misleading because the globular is a member of the Milky Way Galaxy whereas the Small Magellanic Cloud is a satellite of the Milky Way at very much greater distance.

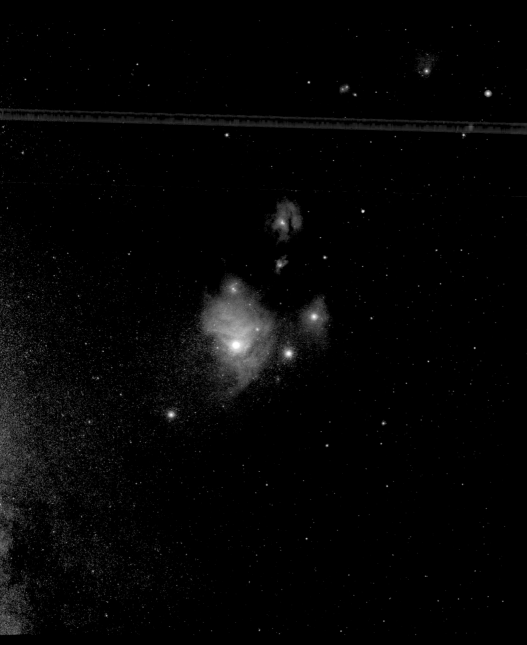

[1] The region of the sky containing the constellation of Scorpius, imaged by the European Space Agency.

[2] This is a 340-million-pixel vista of the central parts of our galactic home, a 34 by 20-degree wide image. It shows the region spanning the sky from the constellation of Sagittarius, the Archer, to Scorpius, the Scorpion. The very colourful Rho Ophiuchi and Antares region features prominently to the right, as well as much darker areas, such as the the Pipe and Snake Nebulae. The dusty lane of our Milky Way runs obliquely through the image, dotted with remarkable bright, reddish nebulae, such as the Lagoon and the Trifid Nebulae, as well as NGC 6357 and NGC 6334. This dark lane also hosts the very centre of our Galaxy, where a supermassive black hole is lurking. ESO/S. Guisard.

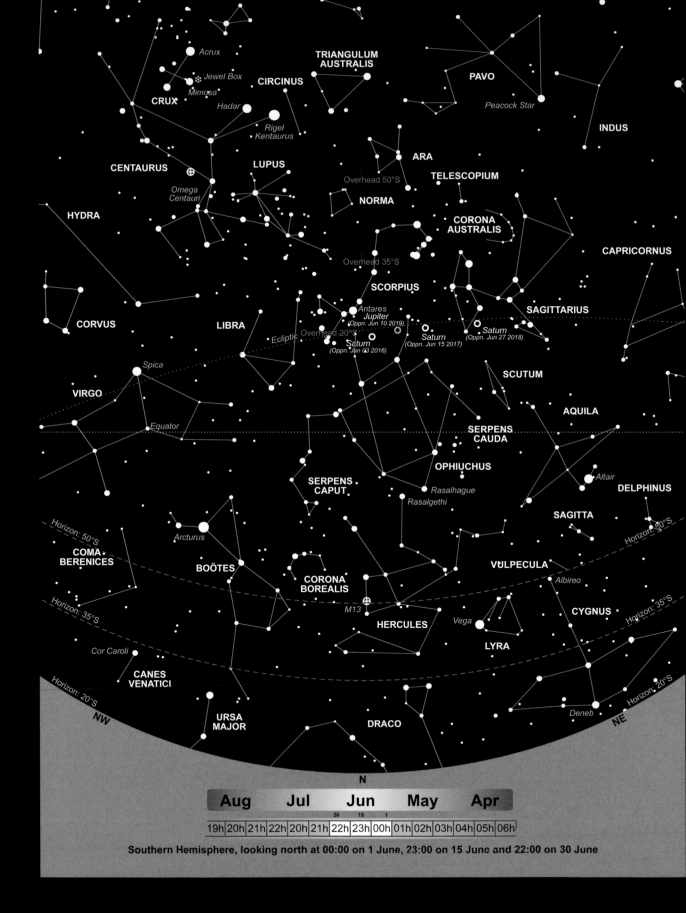

Southern Hemisphere, looking north at 00:00 on 1 June, 23:00 on 15 June and 22:00 on 30 June

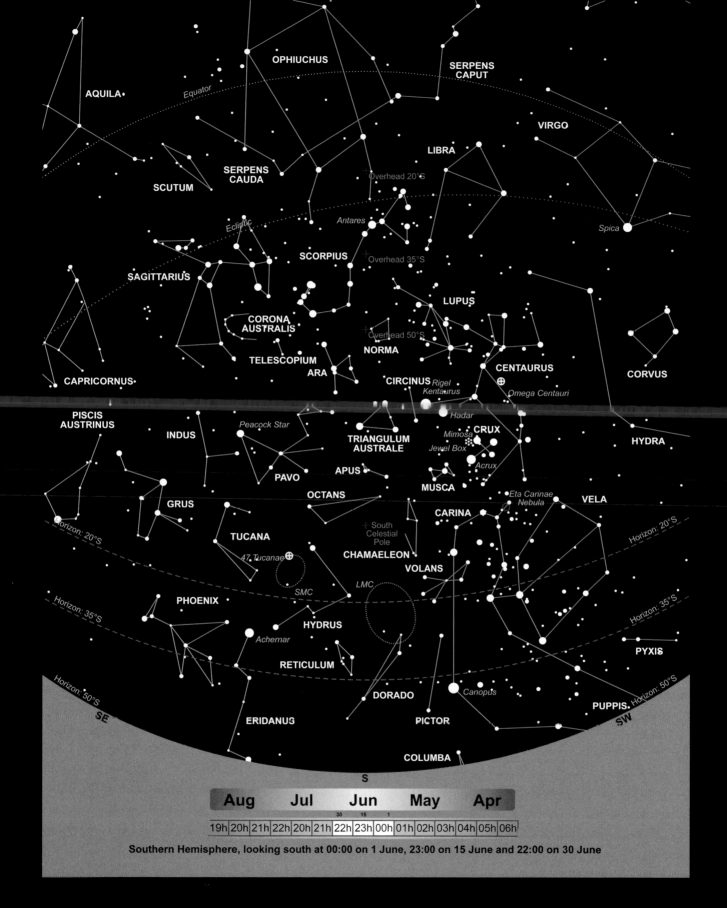

Southern Hemisphere, looking south at 00:00 on 1 June, 23:00 on 15 June and 22:00 on 30 June

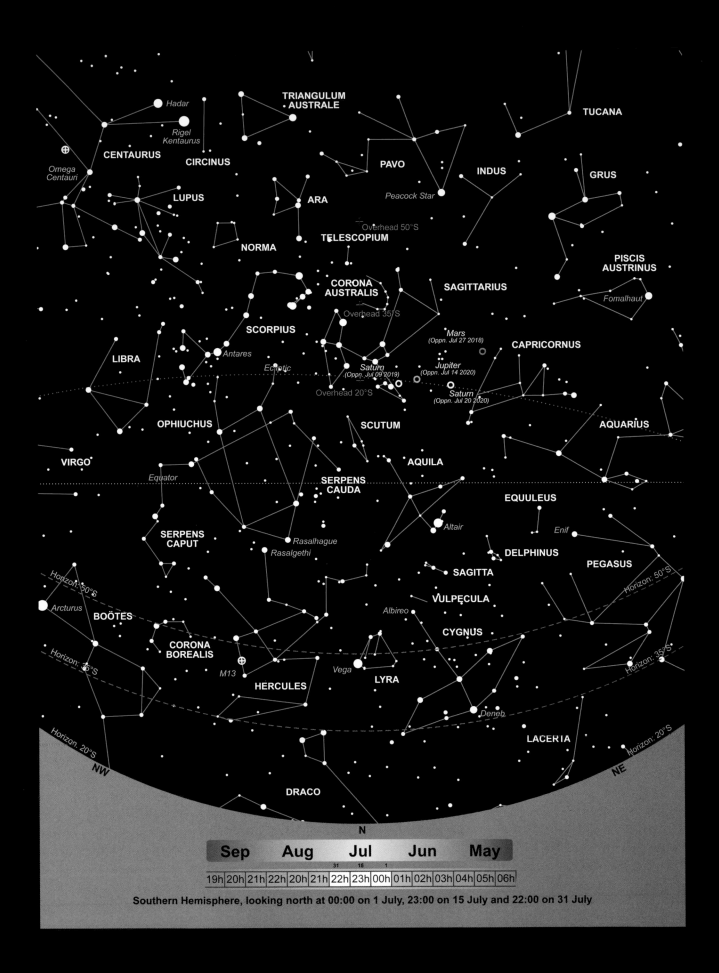

Southern Hemisphere, looking north at 00:00 on 1 July, 23:00 on 15 July and 22:00 on 31 July

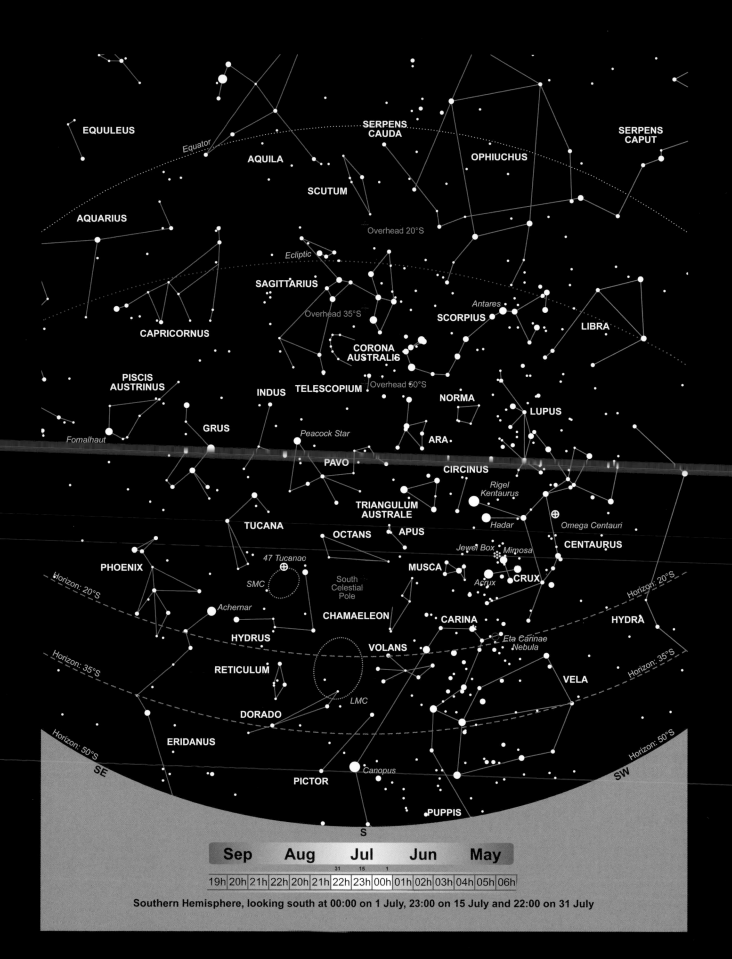

Southern Hemisphere, looking south at 00:00 on 1 July, 23:00 on 15 July and 22:00 on 31 July

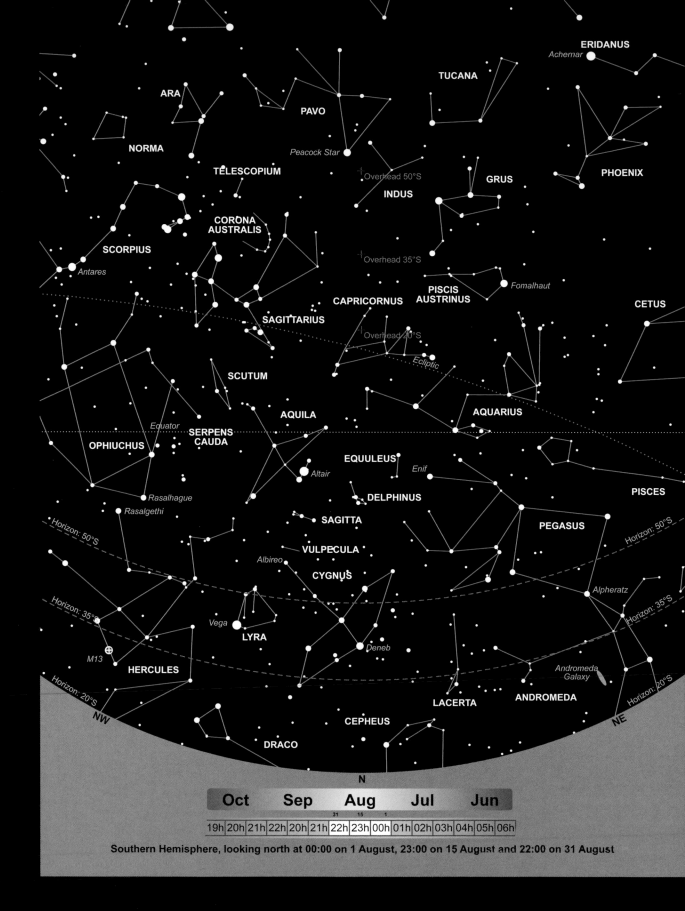

ERIDANUS
Achernar

TUCANA

PAVO

ARA

NORMA

Peacock Star

Overhead 50°S

INDUS

GRUS

PHOENIX

TELESCOPIUM

CORONA
AUSTRALIS

SCORPIUS

Antares

Overhead 35°S

PISCIS
AUSTRINUS

Fomalhaut

CETUS

CAPRICORNUS

Overhead 20°S

SAGITTARIUS

Ecliptic

SCUTUM

AQUILA

AQUARIUS

Equator

OPHIUCHUS

SERPENS
CAUDA

EQUULEUS

Enif

PISCES

Altair

Rasalhague

DELPHINUS

PEGASUS

Alpheratz

Rasalgethi

SAGITTA

Horizon: 50°S

Horizon: 50°S

VULPECULA

Albireo

CYGNUS

Horizon: 35°S

Vega

LYRA

Deneb

Horizon: 35°S

M13

HERCULES

*Andromeda
Galaxy*

Horizon: 20°S

LACERTA

ANDROMEDA

Horizon: 20°S

NW

NE

CEPHEUS

DRACO

N

Oct	Sep	Aug	Jul	Jun

19h	20h	21h	22h	20h	21h	22h	23h	00h	01h	02h	03h	04h	05h	06h

Southern Hemisphere, looking north at 00:00 on 1 August, 23:00 on 15 August and 22:00 on 31 August

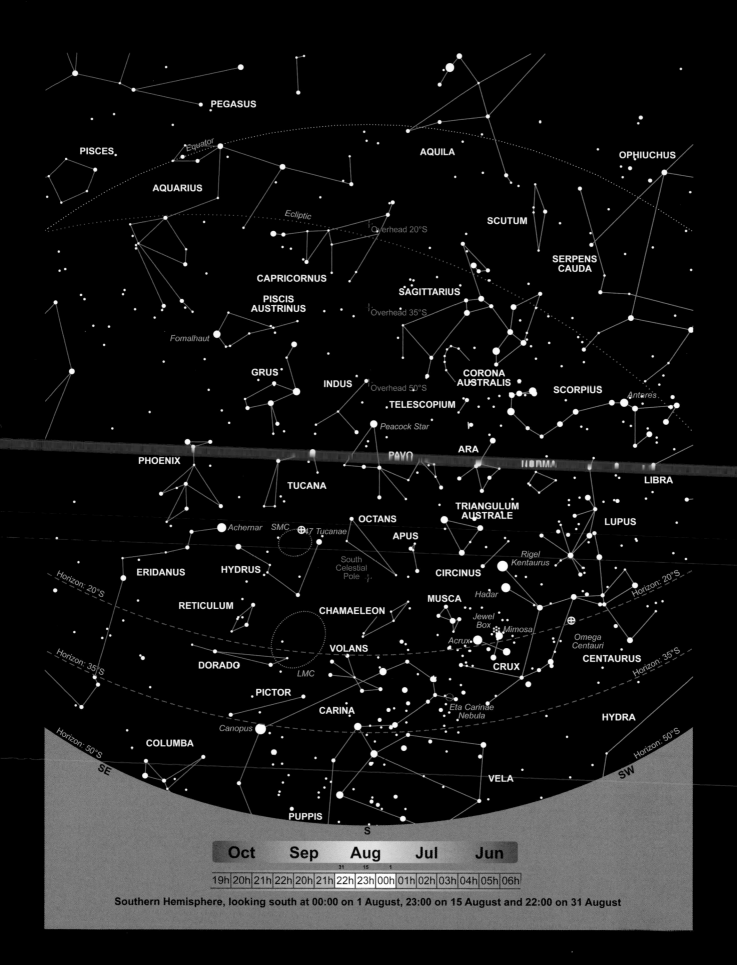

PEGASUS

PISCES

AQUARIUS

Equator

Ecliptic

Overhead 20°S

AQUILA

OPHIUCHUS

SCUTUM

SERPENS
CAUDA

CAPRICORNUS

SAGITTARIUS

PISCIS
AUSTRINUS

Overhead 35°S

Fomalhaut

GRUS

INDUS

Overhead 50°S

CORONA
AUSTRALIS

SCORPIUS

Antares

TELESCOPIUM

Peacock Star

ARA

PAVO

NORMA

LIBRA

PHOENIX

TUCANA

TRIANGULUM
AUSTRALE

LUPUS

Achernar

SMC

47 Tucanae

OCTANS

APUS

CIRCINUS

Rigel
Kentaurus

Horizon: 20°S

ERIDANUS

HYDRUS

South
Celestial
Pole

MUSCA

Hadar

CENTAURUS

Horizon: 20°S

RETICULUM

CHAMAELEON

Jewel
Box

Mimosa

Omega
Centauri

Horizon: 35°S

Acrux

VOLANS

DORADO

LMC

CRUX

Horizon: 35°S

PICTOR

CARINA

Eta Carinae
Nebula

HYDRA

Horizon: 50°S

Canopus

Horizon: 50°S

COLUMBA

SE

VELA

SW

PUPPIS

S

Oct	Sep	Aug	Jul	Jun

19h	20h	21h	22h	20h	21h	22h	23h	00h	01h	02h	03h	04h	05h	06h

31   15   1

Southern Hemisphere, looking south at 00:00 on 1 August, 23:00 on 15 August and 22:00 on 31 August

SPRING (SOUTHERN HEMISPHERE)

In spring we have the Square of Pegasus reasonably high in the northeast but Andromeda is always very low down and it is by no means easy to locate the Andromeda Galaxy M31. Fomalhaut in the Southern Fish is practically at the zenith and one can appreciate how bright it really is. It is a particularly interesting star because it is the centre for a system of planets, and one of these has actually been imaged.

Grus the Crane, adjoins Piscis Austrinus, the Southern Fish – the constellation that has Fomalhaut as its brightest member. Also near the zenith are two large but rather barren constellations, Capricornus, the Sea Goat and Aquarius, the Water Bearer. Aquarius is distinguished by two well-known planetary nebulae, NGC 7009 (Caldwell 55), the Saturn Nebula, and NGC 7293 (Caldwell 63), the Helix Nebula. Curiously, neither of these planetaries are in Messier's catalogue, whereas M73, also in Aquarius, is merely a grouping of four stars.

The Magellanic Clouds

Now let us turn to those superb objects the Magellanic Clouds, which are satellite galaxies of the Milky Way. Both are prominent to the naked eye and any small telescope will show tremendous amounts of detail in them. They are particularly important because they contain objects of all kinds and for most purposes we can regard them as being the same distance from us, though this is not strictly true. The Large Magellanic Cloud is 169 thousand light years from us, and crosses the border between Dorado, the Swordfish, and Mensa, the Table. It contains the magnificent Tarantula Nebula, a gaseous nebula that would cast shadows if it were as close as M42 in the Sword of Orion. In 1987 a supernova blazed up in the Tarantula and became an easy naked eye object.

The Small Magellanic Cloud is 197 thousand light years away, and contains several hundred million stars. It lies in Tucana, and gives the impression of being associated with the globular cluster 47 Tucanae. It contains a number of Cepheid variables, and it was by studying these that Henrietta Swan Leavitt established the vitally important Cepheid period-luminosity law. As they are so prominent to the naked eye, both Magellanic Clouds must have been seen in ancient times, but the first mention of them was by Al-Sufi in 964AD. They were described by the crew members of Ferdinand Magellan's voyage round the world in 1519, hence the name.

Both clouds have been classed as irregular galaxies and in each case there are indications of a rather ill-defined central bar. They are certainly not spirals. The Milky Way galaxy contains a number of other dwarf satellite galaxies of which the closest is the Canis Majoris dwarf. It is hard to detect because it lies behind the plane of the Milky Way and contains an unusual number of red giants. At a mere 25 thousand light years this is the closest galaxy known, and stars have been torn from it by the gravitational pull of our own Galaxy, producing a ring of stars around the Milky Way known as the Monoceros Ring.

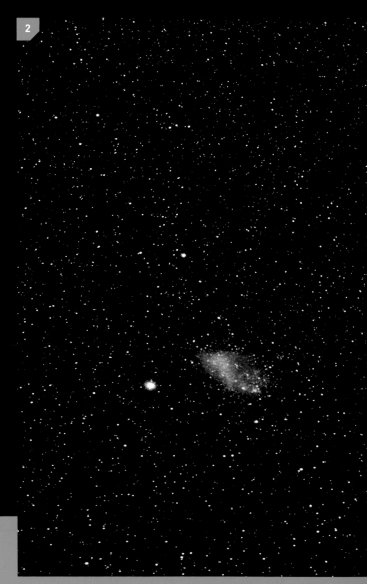

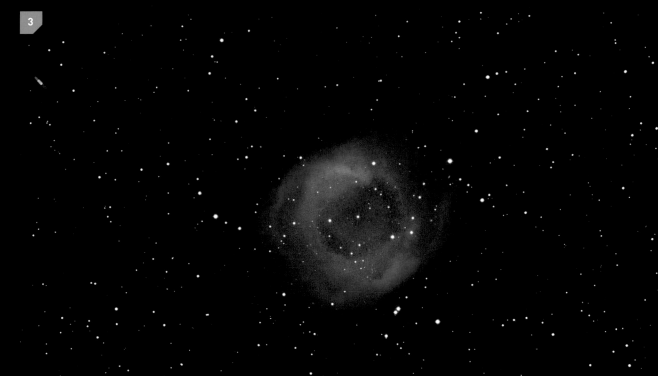

[1] This first light image of the TRAPPIST national telescope at La Silla shows the Tarantula Nebula, located in the Large Magellanic Cloud (LMC) — one of the galaxies closest to us. Also known as 30 Doradus or NGC 2070, the nebula owes its name to the arrangement of bright patches that somewhat resembles the legs of a tarantula. The image was made from data obtained through three filters (B, V and R) and the field of view is about 20 arcminutes across. TRAPPIST/E. Jehin/ESO.

[2] Seen from the southern skies, the Large and Small Magellanic Clouds (the LMC and SMC, respectively) are bright patches in the sky. These two irregular dwarf galaxies, together with our Milky Way Galaxy, belong to the so-called Local Group of galaxies. ESO.

[3] The Helix Nebula (NGC7293), a planetary nebula, imaged with the ESO 3.6-metre telescope on La Silla.2. ESO.

ARA

PAVO

TELESCOPIUM

Peacock Star

CORONA
AUSTRALIS

SAGITTARIUS

CAPRICORNUS

SCUTUM

AQUILA

Equator

EQUULEUS

Altair DELPHINUS

Enif

SAGITTA

VULPECULA

HERCULES

Albireo

CYGNUS

LACERTA

Vega

LYRA

Deneb

Horizon: 50°S

Horizon: 35°S

Horizon: 20°S

NW

47 Tucanae   SMC

HYDRUS

TUCANA

Achernar

ERIDANUS

INDUS

Overhead 50°S

GRUS

PHOENIX

Overhead 35°S

PISCIS
AUSTRINUS

Fomalhaut

Overhead 20°S

CETUS

AQUARIUS

Ecliptic

PISCES

PEGASUS

Horizon: 50°S

Alpheratz

Hamal

ARIES

Horizon: 35°S

Horizon: 20°S

TRIANGULUM

*Andromeda
Galaxy*

ANDROMEDA

NE

CEPHEUS

CASSIOPEIA

N

Nov	Oct	Sep	Aug	Jul

30   15   1

| 19h | 20h | 21h | 22h | 20h | 21h | 22h | 23h | 00h | 01h | 02h | 03h | 04h | 05h | 06h |

Southern Hemisphere, looking north at 00:00 on 1 September, 23:00 on 15 September and 22:00 on 30 September

Southern Hemisphere, looking south at 00:00 on 1 September, 23:00 on 15 September and 22:00 on 30 September

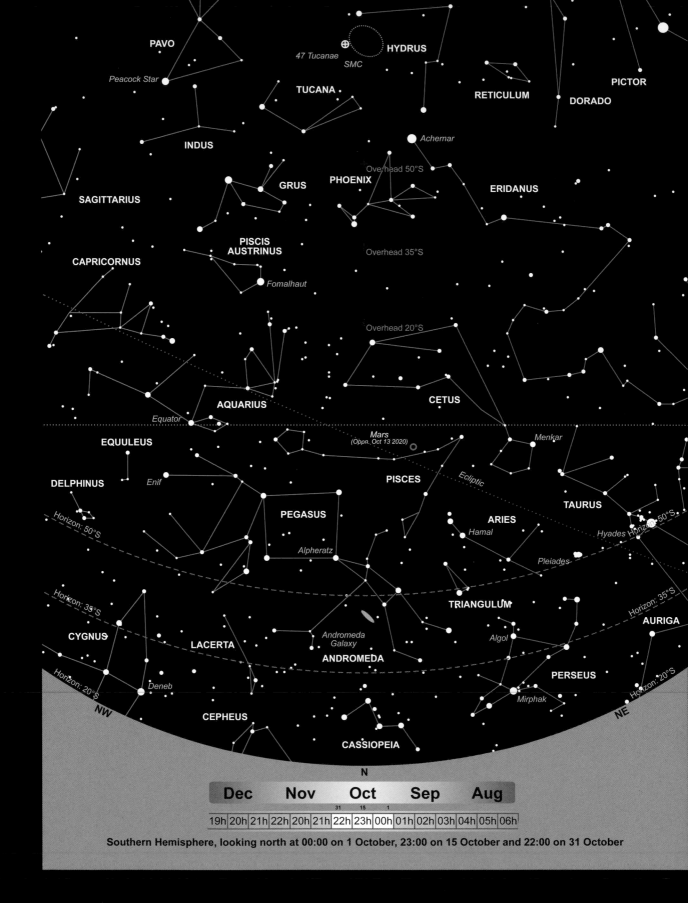

PAVO

47 Tucanae
SMC
HYDRUS

Peacock Star

TUCANA

RETICULUM

PICTOR

DORADO

INDUS

Achernar

Over head 50°S

SAGITTARIUS

GRUS

PHOENIX

ERIDANUS

PISCIS
AUSTRINUS

Overhead 35°S

CAPRICORNUS

Fomalhaut

Overhead 20°S

AQUARIUS

CETUS

Equator

Menkar

EQUULEUS

Mars
(Oppn. Oct 13 2020)

DELPHINUS

Enif

PISCES

Ecliptic

TAURUS

Horizon: 50°S

PEGASUS

ARIES

Hamal

Hyades Horizon: 50°S

Alpheratz

Pleiades

Horizon: 35°S

TRIANGULUM

AURIGA

CYGNUS

LACERTA

Andromeda
Galaxy

Algol

Horizon: 35°S

Horizon: 20°S

ANDROMEDA

PERSEUS

Horizon: 20°S

Deneb

Mirphak

NW

NE

CEPHEUS

CASSIOPEIA

N

Dec	Nov	Oct	Sep	Aug

19h	20h	21h	22h	20h	21h	22h	23h	00h	01h	02h	03h	04h	05h	06h

Southern Hemisphere, looking north at 00:00 on 1 October, 23:00 on 15 October and 22:00 on 31 October

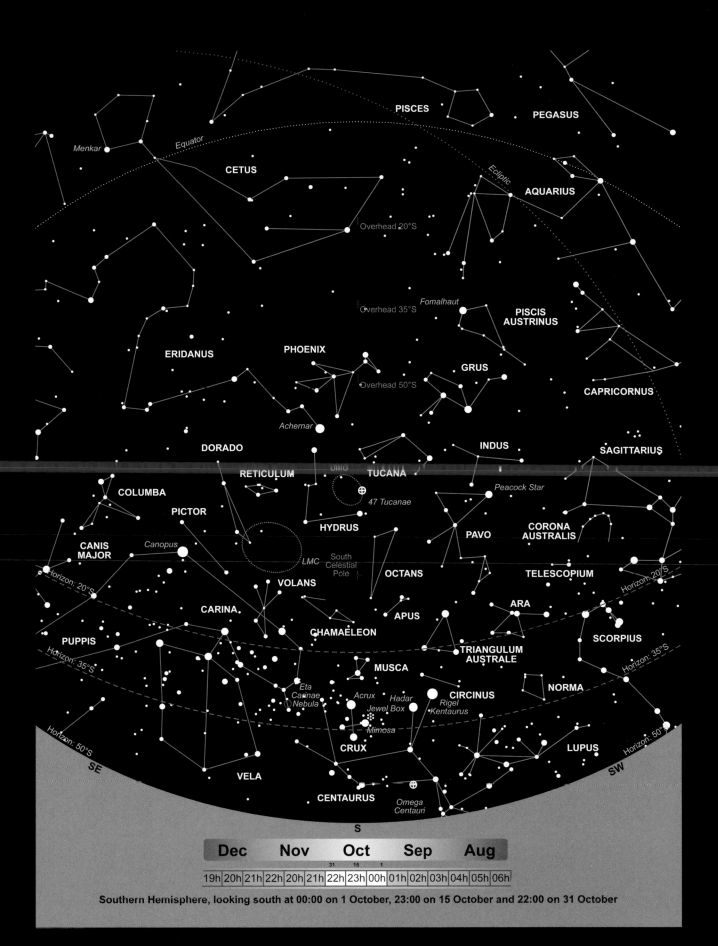

PISCES

PEGASUS

Menkar *Equator*

CETUS

Ecliptic

AQUARIUS

Overhead 20°S

Overhead 35°S *Fomalhaut*

PISCIS
AUSTRINUS

ERIDANUS

PHOENIX

GRUS

CAPRICORNUS

Overhead 50°S

Achernar

DORADO

INDUS

SAGITTARIUS

RETICULUM

TUCANA

Peacock Star

COLUMBA

47 Tucanae

PICTOR

HYDRUS

PAVO

CORONA
AUSTRALIS

CANIS
MAJOR

Canopus

LMC

South
Celestial
Pole

OCTANS

TELESCOPIUM

Horizon: 20°S

VOLANS

APUS

ARA

SCORPIUS

Horizon: 20°S

CARINA

CHAMAELEON

PUPPIS

MUSCA

TRIANGULUM
AUSTRALE

NORMA

Horizon: 35°S

*Eta
Carinae
Nebula*

Acrux *Hadar*

CIRCINUS

Horizon: 35°S

Jewel Box

*Rigel
Kentaurus*

Horizon: 50°S

Mimosa

CRUX

LUPUS

Horizon: 50°S

SE

VELA

SW

CENTAURUS

*Omega
Centauri*

S

Dec   Nov   Oct   Sep   Aug

31   15   1

19h|20h|21h|22h|20h|21h|22h|23h|00h|01h|02h|03h|04h|05h|06h

Southern Hemisphere, looking south at 00:00 on 1 October, 23:00 on 15 October and 22:00 on 31 October

Southern Hemisphere, looking north at 00:00 on 1 November, 23:00 on 15 November and 22:00 on 30 November

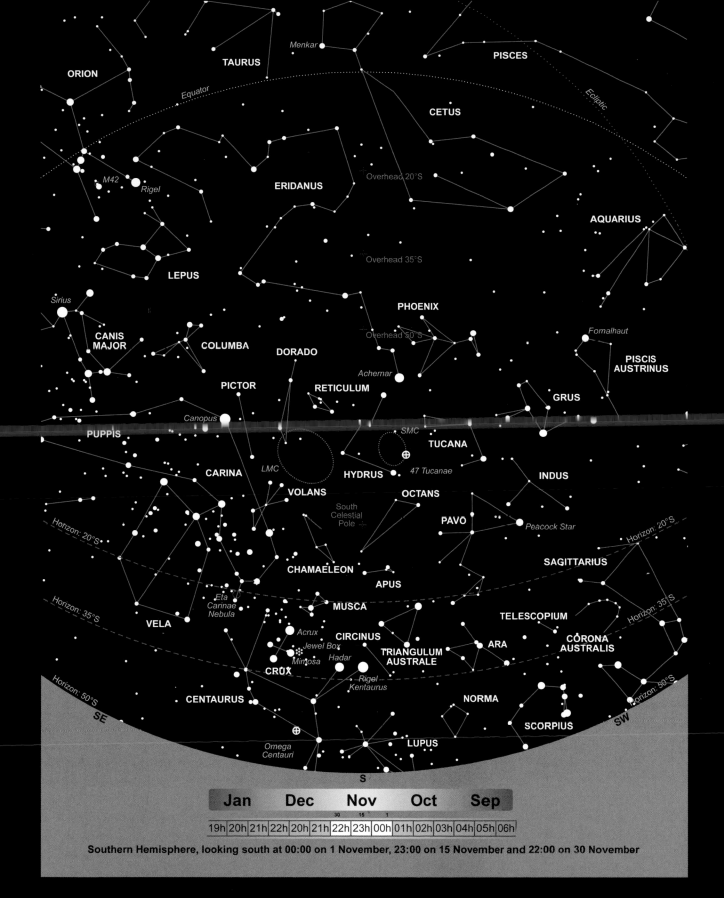

ORION
TAURUS
Menkar
PISCES
CETUS
Equator
Ecliptic
M42
Rigel
ERIDANUS
Overhead 20°S
AQUARIUS
LEPUS
Overhead 35°S
Sirius
PHOENIX
Fomalhaut
CANIS
MAJOR
COLUMBA
DORADO
Overhead 50°S
PISCIS
AUSTRINUS
Achernar
PICTOR
RETICULUM
GRUS
Canopus
SMC
PUPPIS
TUCANA
INDUS
CARINA
LMC
HYDRUS
47 Tucanae
VOLANS
OCTANS
*South
Celestial
Pole +*
PAVO
Peacock Star
CHAMAELEON
APUS
SAGITTARIUS
Horizon: 20°S
Horizon: 20°S
MUSCA
TELESCOPIUM
Horizon: 35°S
*Eta
Carinae
Nebula*
CIRCINUS
ARA
CORONA
AUSTRALIS
Horizon: 35°S
VELA
Acrux
Jewel Box
Mimosa
Hadar
TRIANGULUM
AUSTRALE
Horizon: 50°S
CRUX
*Rigel
Kentaurus*
NORMA
CENTAURUS
SCORPIUS
SW
Horizon: 50°S
SE
*Omega
Centauri*
LUPUS
S

Jan	Dec	Nov	Oct	Sep

30 15 1

| 19h | 20h | 21h | 22h | 20h | 21h | 22h | 23h | 00h | 01h | 02h | 03h | 04h | 05h | 06h |

Southern Hemisphere, looking south at 00:00 on 1 November, 23:00 on 15 November and 22:00 on 30 November

SUMMER (SOUTHERN HEMISPHERE)

It has to be said that the stars in the far south of the sky are much more distinctive than those of the far north. From New Zealand, for example, you will look in vain for the Great Bear, although Orion, crossed by the celestial equator, is on view. The most famous constellation is the brilliant Southern Cross. In summer it is high up in the sky. It is probably the best guide to the southern pole of the sky which is not itself marked by any bright star.

To find the pole, the best method is to follow a line along the longer axis of the Cross until you come to a really brilliant star, Achenar in Eridanus. The pole lies roughly midway between Achenar and the Cross. The trouble here is that when either of these two locators is low, they may be lost to the haze close to the horizon, making it hard to find the pole in this way all the time. The Cross is not circumpolar north of Sydney or Cape Town.

The Pole Star

The pole actually lies in the constellation of Octans, which is very barren indeed. The actual pole star, Sigma Octantis, is only of magnitude 5.5 and is never really an easy naked eye object, it is also hidden by the slightest haze. Moreover, there is nothing distinctive anywhere near it, so you will have to search carefully.

In summer in the southern hemisphere Orion is prominent

with Sirius much higher in the sky than it ever appears in Britain. This means it twinkles and flashes much less. Also high up is the splendid constellation of Carina, the Keel of the ship Argo, a constellation so large that the International Astronomical Union broke it up into Carina, the Keel, Puppis, the Poop, and Vela, the Sails.

The brightest star in Carina is Canopus, which is the most brilliant star in the sky apart from Sirius. It has an F-type spectrum meaning the star is yellow-white in colour, though most see it as white. It is an exceptionally powerful star, tens of thousands of times more luminous than our Sun.

Eta Carinae

Apart from Canopus, Carina contains many remarkable objects, one of these is the erratic variable Eta Carinae, which is a remote hypergiant. At one time, around 1840, it was the brightest star in the sky apart from Sirius, but today it is only just visible with the naked eye. This does not mean its luminosity

has decreased; it's just that most radiation is at infrared wavelengths. It is a highly unstable star, and in the foreseeable future will flare up as a supernova. This may happen today, tomorrow, or in a million years or more, but happen it will.

The Southern Cross

The Southern Cross, Crux, contains the lovely double star Alpha Crucis. Here too is the Jewel Box, a lovely cluster with one prominent red star.

Crux is more or less surrounded by Centaurus, the Centaur. Here we have two first magnitude stars Alpha Centauri or Rigel Kentaurus and Beta Centauri, otherwise known as Agena or Hadar. They lie side by side in the sky but are not connected. Agena is a remote giant star hundreds of light years from us while Alpha Centauri is a member of a triple star sytem, one member of which, Proxima Centauri, is the nearest star in the sky. Its distance is 4.2 light years.

In the northern part of the sky Orion is in evidence.

[1] This wide-field image obtained with an Hasselblad 2000 FC camera by Claus Madsen (ESO) shows a region around the Southern Cross, seen in the right of the image (Kodak Ektachrome 200, 70 min exposure time). Alpha Centauri is the bright yellowish star seen at the middle left, one of the "Pointers" to the star at the top of the Southern Cross. ESO, Claus Madsen

[2] This spectacular panoramic view combines a new image of the field around the Wolf–Rayet star WR 22 in the Carina Nebula (right) with an earlier picture of the region around the unique star Eta Carinae in the heart of the nebula (left). The picture was created from images taken with the Wide Field Imager on the MPG/ESO 2.2-metre telescope at ESO's La Silla Observatory in Chile.

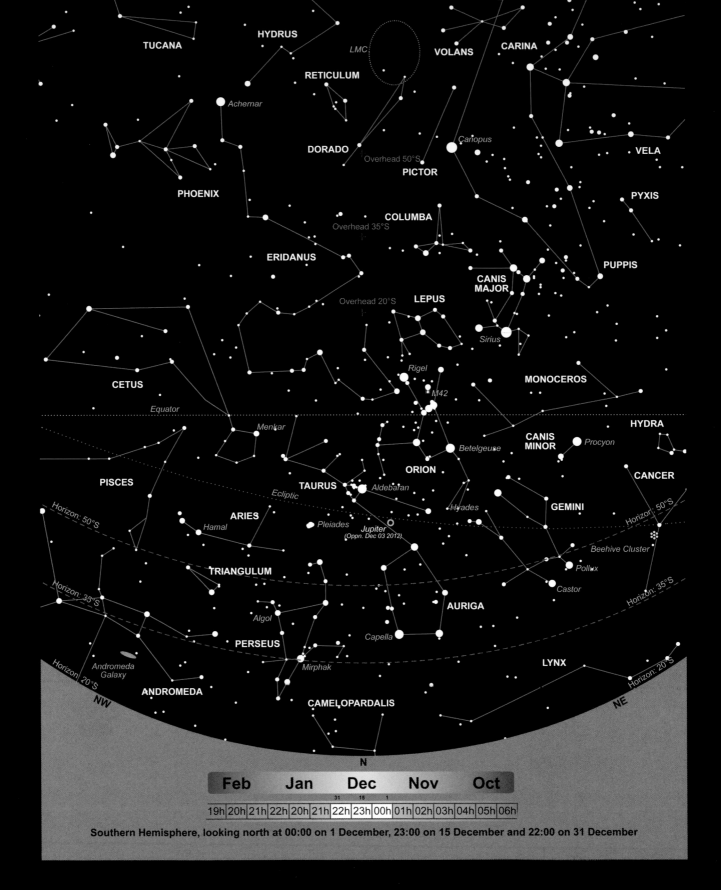

TUCANA

HYDRUS

LMC

VOLANS

CARINA

RETICULUM

Achernar

DORADO

Canopus

VELA

Overhead 50°S

PICTOR

PHOENIX

PYXIS

COLUMBA

Overhead 35°S

PUPPIS

ERIDANUS

LEPUS

CANIS MAJOR

Sirius

MONOCEROS

Overhead 20°S

Rigel

CETUS

M42

HYDRA

Equator

Menkar

Betelgeuse

CANIS MINOR

Procyon

PISCES

ORION

CANCER

Aldebaran

Horizon: 50°S

ARIES

TAURUS

Hyades

GEMINI

Horizon: 50°S

Hamal

Pleiades

Ecliptic

Jupiter
(Oppn. Dec 03 2012)

Beehive Cluster

TRIANGULUM

Pollux

Horizon: 35°S

Horizon: 35°S

Algol

Castor

AURIGA

PERSEUS

Capella

Horizon: 20°S

LYNX

Horizon: 20°S

Andromeda Galaxy

Mirphak

NW

ANDROMEDA

CAMELOPARDALIS

NE

N

Feb	Jan	Dec	Nov	Oct

31   15   1

| 19h | 20h | 21h | 22h | 20h | 21h | 22h | 23h | 00h | 01h | 02h | 03h | 04h | 05h | 06h |

Southern Hemisphere, looking north at 00:00 on 1 December, 23:00 on 15 December and 22:00 on 31 December

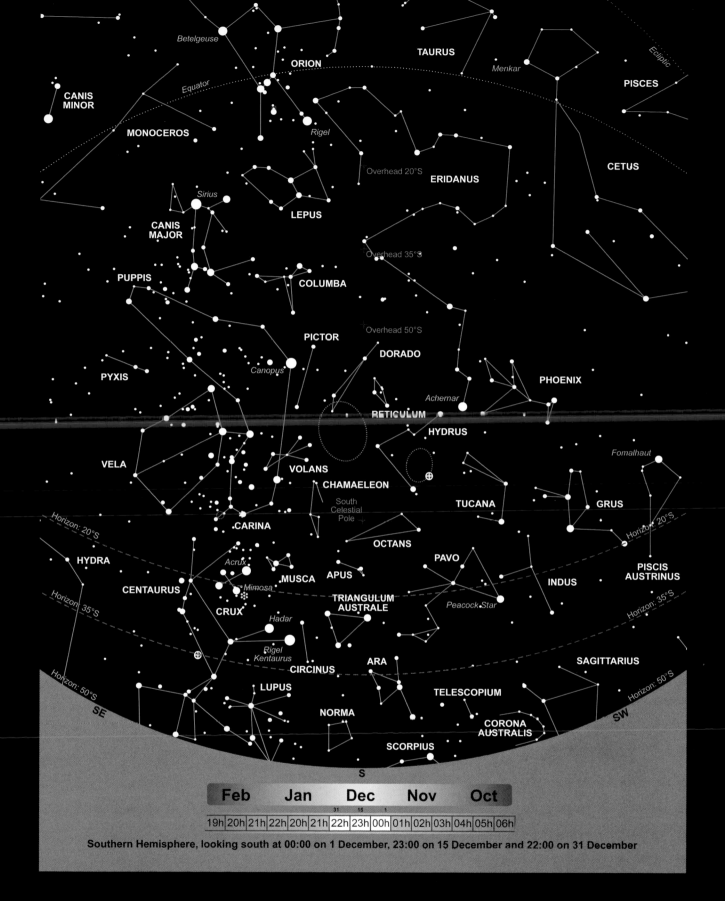

Southern Hemisphere, looking south at 00:00 on 1 December, 23:00 on 15 December and 22:00 on 31 December

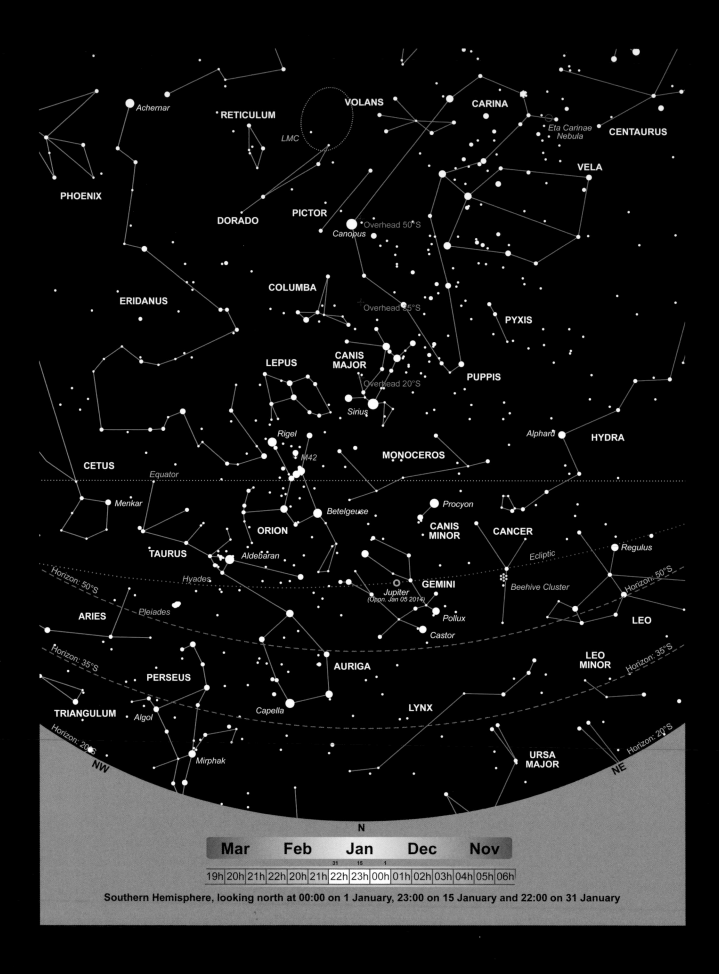

PHOENIX

Achernar

RETICULUM

VOLANS

LMC

CARINA

Eta Carinae
Nebula

CENTAURUS

VELA

DORADO

PICTOR

Overhead 50°S

Canopus

COLUMBA

Overhead 35°S

PYXIS

ERIDANUS

LEPUS

CANIS
MAJOR

Overhead 20°S

PUPPIS

Sirius

Rigel

M42

MONOCEROS

Alphard

HYDRA

CETUS

Equator

Menkar

Betelgeuse

Procyon

CANIS
MINOR

CANCER

Ecliptic

Regulus

ORION

Horizon: 50°S

TAURUS

Aldebaran

Hyades

GEMINI

Jupiter
(Oppn. Jan 05 2014)

Beehive Cluster

Horizon: 50°S

ARIES

Pleiades

Pollux

LEO

Horizon: 35°S

Castor

AURIGA

LEO
MINOR

Horizon: 35°S

PERSEUS

Capella

LYNX

TRIANGULUM

Algol

Horizon: 20°S

Mirphak

URSA
MAJOR

Horizon: 20°S

NW

NE

N

Mar	Feb	Jan	Dec	Nov

31   15   1

| 19h | 20h | 21h | 22h | 20h | 21h | 22h | 23h | 00h | 01h | 02h | 03h | 04h | 05h | 06h |

Southern Hemisphere, looking north at 00:00 on 1 January, 23:00 on 15 January and 22:00 on 31 January

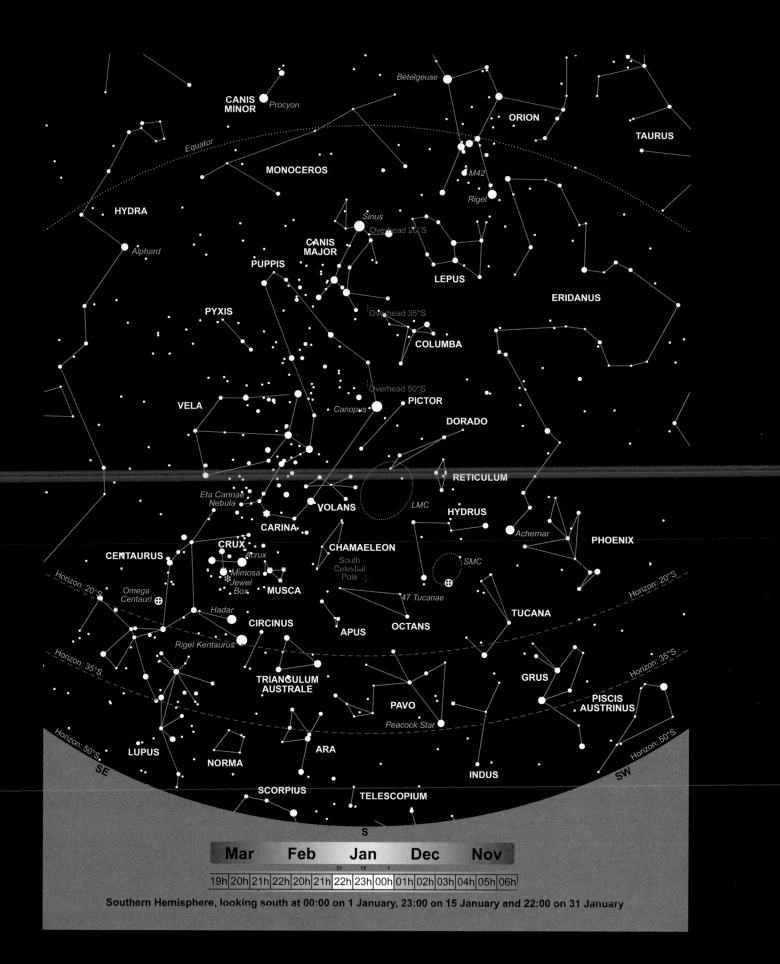

Southern Hemisphere, looking south at 00:00 on 1 January, 23:00 on 15 January and 22:00 on 31 January

RETICULUM

LMC

VOLANS

CARINA

CRUX *Acrux*
✳ *Jewel Box*
× *Mimosa*

DORADO

PICTOR

Eta Carinae Nebula

⊕ *Omega Centauri*

CENTAURUS

Canopus

+ Overhead 50°S

VELA

COLUMBA

ERIDANUS

+ Overhead 35°S

PYXIS

CORVUS

LEPUS

CANIS MAJOR

PUPPIS

+ Overhead 20°S

HYDRA

CRATER

Sirius

Rigel

M42

MONOCEROS

Alphard

Equator

CANIS MINOR

Procyon

Ecliptic

VIRGO

ORION

Betelgeuse

Regulus

Denebola

○ *Jupiter*
(Oppn. Feb 06 2015)

GEMINI

❀ *Beehive Cluster*

Horizon: 50°S

Aldebaran

LEO

COMA BERENICES

Hyades

Pollux

CANCER

Horizon: 35°S

TAURUS

Castor

LEO MINOR

AURIGA

Horizon: 35°S

Horizon: 20°S

LYNX

CANES VENATICI

Cor Caroli

Horizon: 20°S

URSA MAJOR

Capella

NW

Plough

NE

N

Apr	Mar	Feb	Jan	Dec

28 15 1

19h	20h	21h	22h	20h	21h	22h	23h	00h	01h	02h	03h	04h	05h	06h

Southern Hemisphere, looking north at 00:00 on 1 February, 23:00 on 15 February and 22:00 on 28 February

Southern Hemisphere, looking south at 00:00 on 1 February, 23:00 on 15 February and 22:00 on 28 February

1

Autumn skies are brilliant in the southern hemisphere. Centaurus, with the Southern Cross is very close to the zenith. Adjoining the Centaur we have first of all the Scorpion, with its long line of bright stars, including Antares, and then in the east are the star clouds of Sagittarius. On a dark autumn night these clouds are truly superb, and the shadows they can cast from really dark sky sites are quite distinctive.

Of course these clouds mask our view of the centre of the Galaxy, an intensely interesting region which has to be studied by indirect means. Sweeping round Sagittarius with a pair of binoculars or a wide-field telescope is very rewarding, and this is arguably the richest area in the whole of the sky. Very low in the west, Achenar can probably be made out, but it is not far above the horizon and any mist will obscure it. This makes the

location of the south celestial pole considerably more difficult. In the western part of the sky Carina dominates with Canopus still at a respectable altitude, though we have lost Sirius.

This is a particularly good time to look at the two best double stars in the sky – Alpha Centauri and Alpha Crucis. Do not forget the erratic variable Eta Carinae, which could flare up again at any moment and is always worth following.

2

One prominent feature of the autumn sky is Corvus, the Crow or Raven. Its four main stars are little brighter than the third magnitude but are so distinctive they are impossible to overlook. Another constellation, Libra, the Scales, lies between the constellations of Virgo and Scorpius.

Zubeneschamali

One star in Libra, Beta Librae, with the extraordinary name of Zubeneschamali, is said to be the only single naked eye star that is green in colour! PM has looked very hard at Beta Librae with the naked eye, binoculars and telescopes, and once or twice has suspected a greenish tinge, but this is possibly due to wishful thinking! The only obviously green stars are the companions to red giants such as Antares and Alpha Herculis (Rasalgethi), and here the greenness is almost certainly due to nothing more than colour contrast.

Leo, the Lion

Turn now to the northern aspect. Leo is sitting in the west,

with Arcturus and Spica high up. Virgo is a large constellation in the area between here and Leo, and is very rich in galaxies. Arcturus is prominent in the northern part of the sky, and one can appreciate how brilliant it really is. It is one of only three stars above zero magnitude – the other two are Sirius and Canopus.

The Great Globular Clusters

Also on view are Ophiuchus and Hercules. Find the globular cluster in Hercules and compare it with Omega Centauri – it is obvious that Omega Centauri is much the more populous of the two and is believed to have a black hole at its centre. In fact, it is so massive that there have been suspicions it was once an independent galaxy which was swept up by the stars of Centaurus and this may well be so.

Two more constellations – Triangulum Australe, the Southern Triangle, is one of the few constellations that looks like the object it is supposed to represent. Its brightest star, Alpha, is of the second magnitude and somewhat reddish.

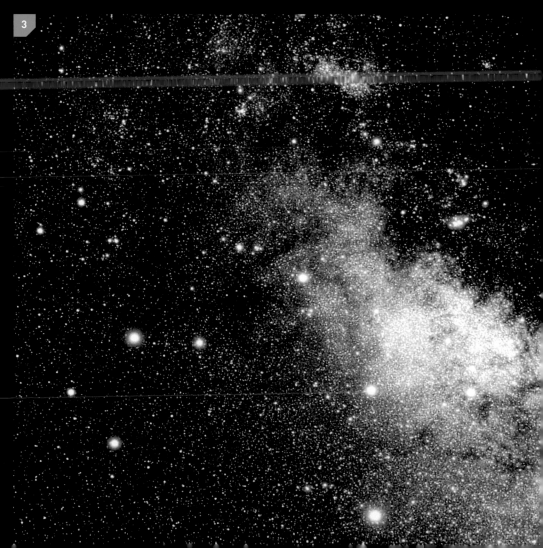

1] The globular cluster Omega Centauri seen in all its splendour in this image by Pete Lawrence.

2] Pete Lawrence captures the hundreds of thousands of stars moving about in the globular cluster M13 in the constellation Hercules.

3] The constellation Sagittarius showing some of the brighter stars in the shape of a teapot in the lower part of the image, with the rich star clouds above. ESA, NASA, Akira Fujii.

3

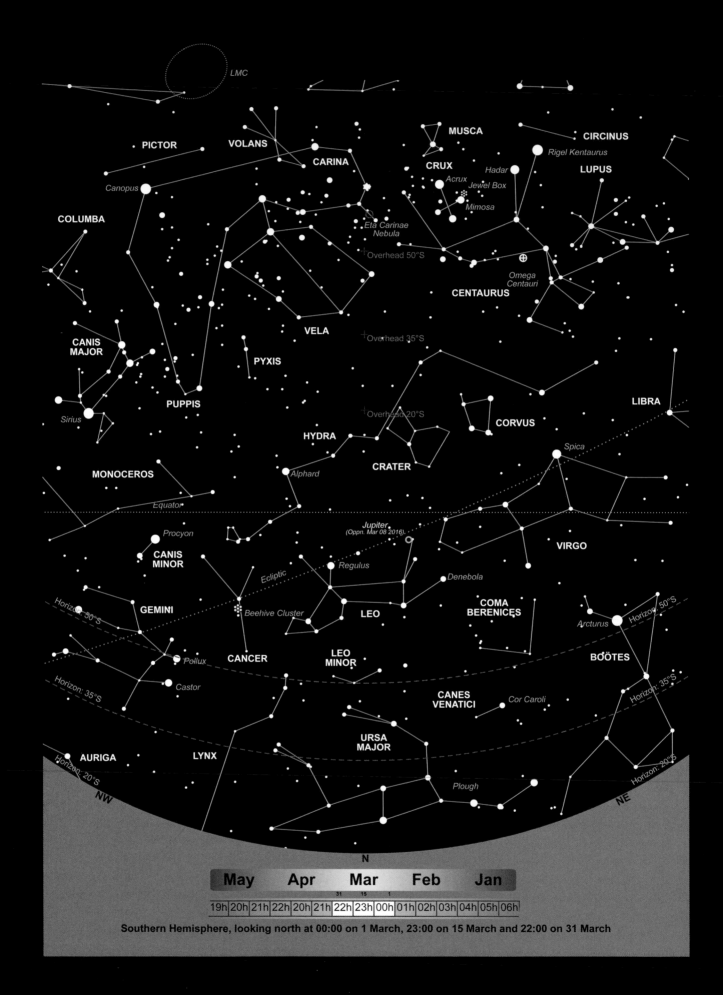

LMC

PICTOR   VOLANS   CARINA   MUSCA   CIRCINUS

Canopus   Eta Carinae Nebula   CRUX   Acrux   Jewel Box   Mimosa   Hadar   Rigel Kentaurus   LUPUS

COLUMBA

Overhead 50°S   Omega Centauri   CENTAURUS

CANIS MAJOR   VELA   Overhead 35°S

Sirius   PYXIS   PUPPIS

LIBRA

MONOCEROS   HYDRA   CRATER   Overhead 20°S   CORVUS   Spica

Alphard   CRATER   VIRGO

Equator   Jupiter (Oppn. Mar 08 2016)

Procyon   CANIS MINOR   Regulus   Denebola   COMA BERENICES   Arcturus   Horizon: 50°S

Ecliptic   Beehive Cluster

Horizon: 50°S   GEMINI   LEO   LEO MINOR   BOÖTES   Horizon: 35°S

Pollux   CANCER   CANES VENATICI   Cor Caroli   Horizon: 20°S

Horizon: 35°S   Castor

AURIGA   LYNX   URSA MAJOR

Horizon: 20°S   Plough

NW   N   NE

May	Apr	Mar	Feb	Jan

31 15 1

19h	20h	21h	22h	20h	21h	22h	23h	00h	01h	02h	03h	04h	05h	06h

Southern Hemisphere, looking north at 00:00 on 1 March, 23:00 on 15 March and 22:00 on 31 March

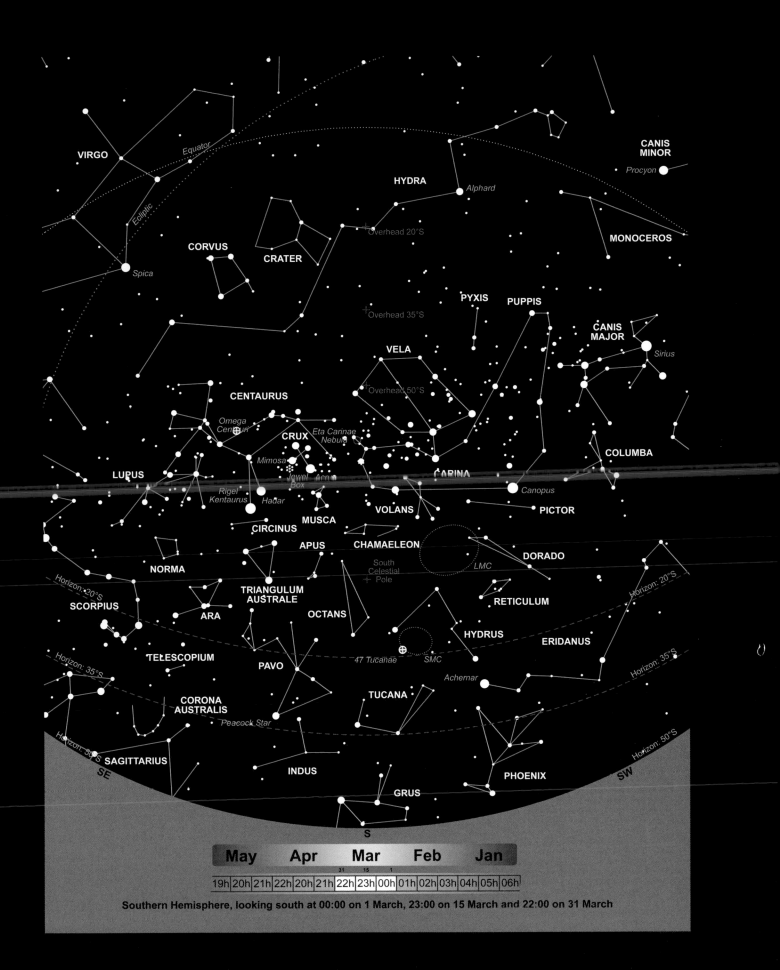

Southern Hemisphere, looking south at 00:00 on 1 March, 23:00 on 15 March and 22:00 on 31 March

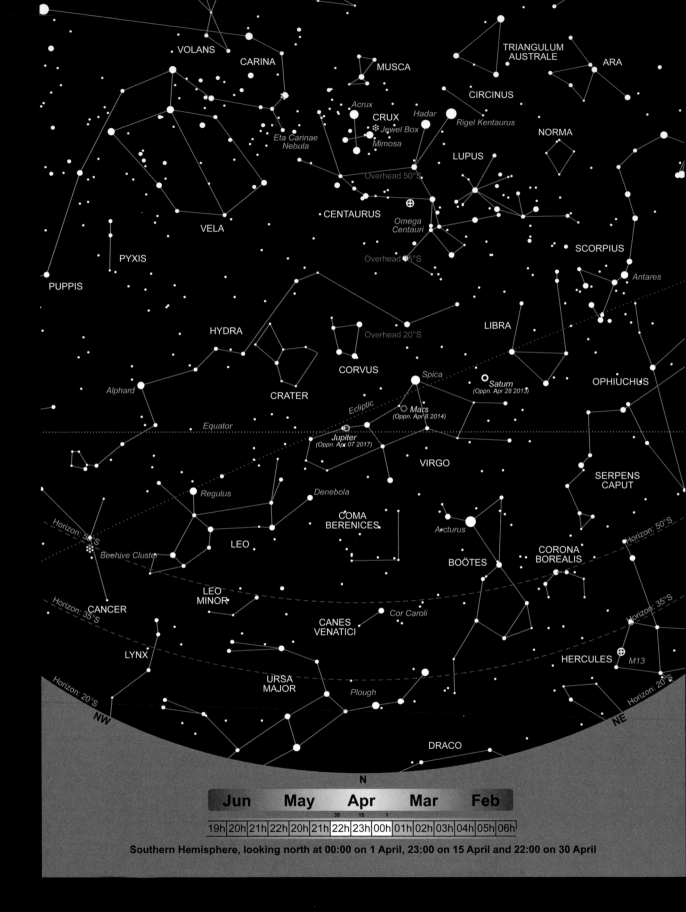

Southern Hemisphere, looking north at 00:00 on 1 April, 23:00 on 15 April and 22:00 on 30 April

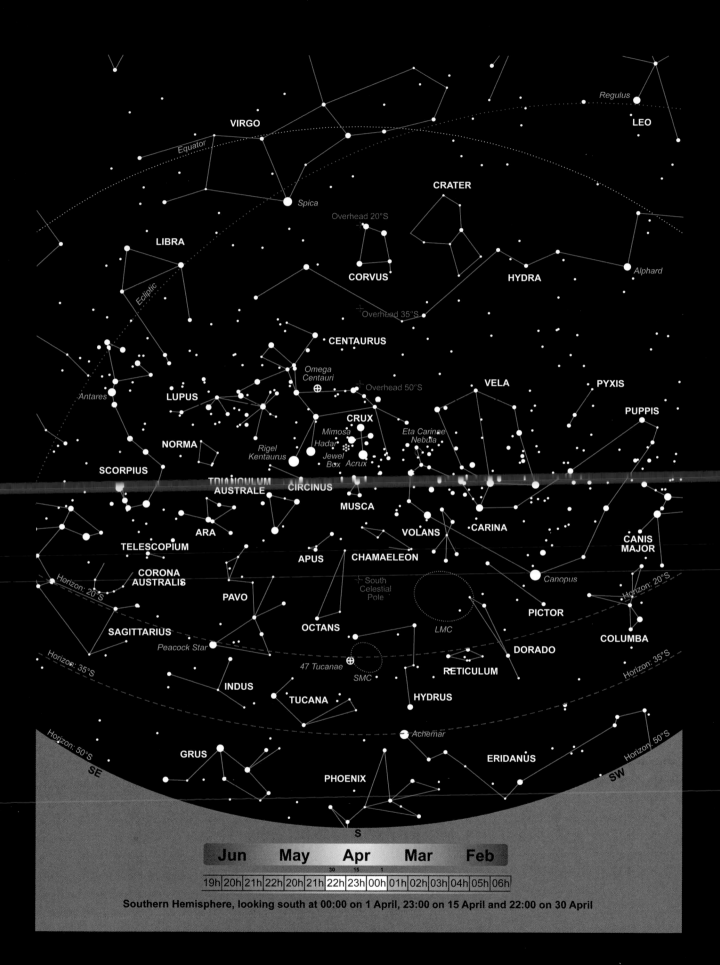

Southern Hemisphere, looking south at 00:00 on 1 April, 23:00 on 15 April and 22:00 on 30 April

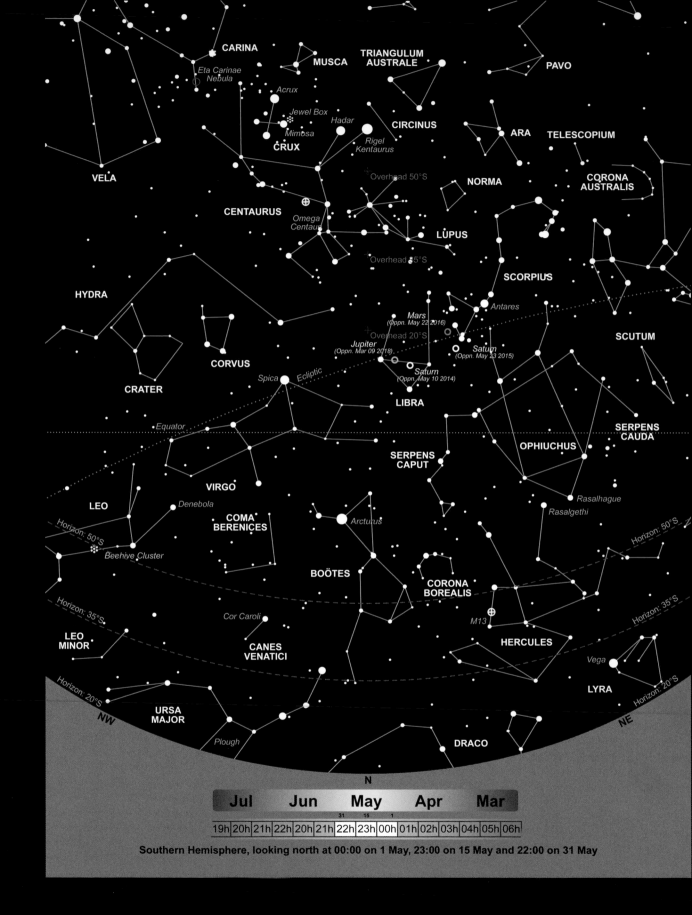

CARINA

MUSCA

TRIANGULUM
AUSTRALE

PAVO

*Eta Carinae
Nebula*

Acrux

Jewel Box

Mimosa

Hadar

CIRCINUS

ARA

TELESCOPIUM

CRUX

*Rigel
Kentaurus*

VELA

Overhead 50°S

NORMA

CORONA
AUSTRALIS

CENTAURUS

*Omega
Centauri*

LUPUS

Overhead 35°S

SCORPIUS

HYDRA

Antares

SCUTUM

*Mars
(Oppn. May 22 2016)*

Overhead 20°S

*Saturn
(Oppn. May 23 2015)*

*Jupiter
(Oppn. Mar 09 2018)*

CORVUS

*Saturn
(Oppn. May 10 2014)*

CRATER

Spica *Ecliptic*

LIBRA

SERPENS
CAUDA

Equator

SERPENS
CAPUT

OPHIUCHUS

VIRGO

LEO

Denebola

COMA
BERENICES

Arcturus

Rasalhague

Rasalgethi

Horizon: 50°S

Beehive Cluster

BOÖTES

CORONA
BOREALIS

Horizon: 50°S

LEO
MINOR

Cor Caroli

M13

Horizon: 35°S

CANES
VENATICI

HERCULES

Vega

Horizon: 20°S

URSA
MAJOR

LYRA

Horizon: 20°S

NW

DRACO

NE

Plough

N

Jul	Jun	May	Apr	Mar

31 15 1

19h	20h	21h	22h	20h	21h	22h	23h	00h	01h	02h	03h	04h	05h	06h

Southern Hemisphere, looking north at 00:00 on 1 May, 23:00 on 15 May and 22:00 on 31 May

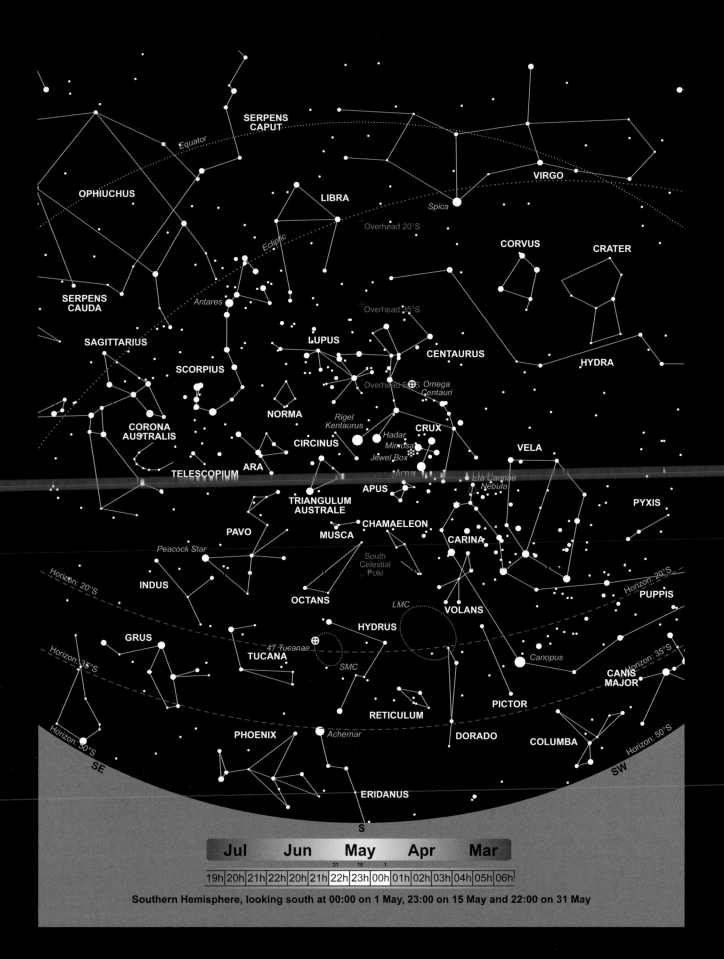

Southern Hemisphere, looking south at 00:00 on 1 May, 23:00 on 15 May and 22:00 on 31 May

GLOSSARY

Absolute magnitude. The apparent magnitude that a star would have if it could be observed from a standard distance of 10 parsecs (32.6 light years).

Absolute zero. The coldest possible temperature: –273.16 °C. Achromatic object-glass. An object-glass which has been corrected so as to eliminate chromatic aberration or false colour as much as possible.

Airglow. The light produced and emitted by the Earth's atmosphere (excluding meteor trains, thermal radiation, lightning and auroræ). Albedo. The reflecting power of a planet or other non-luminous body. The Moon is a poor reflector; its albedo is a mere 7% on average. Altazimuth mounting for a telescope. A mounting on which the telescope may swing freely in any direction.

Altitude. The angular distance of a celestial body above the horizon.

Analemma. The figure-of-eight shape resulting if the Sun's position in the sky is recorded at the same time of day throughout the year.

Ångström unit. One hundred-millionth part of a centimetre.

Aphelion. The furthest distance of a planet or other body from the Sun in its orbit.

Apogee. The furthest point of the Moon from the Earth in its orbit

Apparent magnitude. The apparent brightness of a celestial body. The lower the magnitude, the brighter the object: thus the Sun is approximately –27, the Pole Star +2, and the faintest stars detect able by modern techniques around +30.

Asterism. A pattern of stars which does not rank as a separate constellation.

Asteroids. One of the names for the minor planet swarm.

Astronomical unit. The mean distance between the Earth and the Sun. It is equal to 149 598 500 km.

Aurora. Auroræ are 'polar lights'; Aurora Borealis (northern) and Aurora Australis (southern). They occur in the Earth's upper atmosphere, and are caused by charged particles emitted by the Sun.

Azimuth. The bearing of an object in the sky, measured from north (0°) through east, south and west.

Bailly's beads. Brilliant points seen along the edge of the Moon just before and just after a total solar eclipse. They are caused by the sunlight shining through valleys at the Moon's limb.

Barycentre. The centre of gravity of the Earth Moon system. Because the Earth is 81 times as massive as the Moon, the barycentre lies well inside the Earth's globe.

Binary star. A stellar system made up of two stars, genuinely associated, and moving round their common centre of gravity. The revolution periods range from millions of years for very widely separated visual pairs down to less than half an hour for pairs in which the components are almost in contact with each other. With very close pairs, the components cannot be seen separately, but may be detected by spectroscopic methods.

Black hole. A region round a very small, very massive collapsed star from which not even light can escape.

Caldera (pl calderæ). A large depression, usually found at the summit of a shield volcano, due to the withdrawal of magma from below.

Cassegrain reflector. A reflecting telescope in which the secondary mirror is convex; the light is passed back through a hole in the main mirror. Its main advantage is that it is more compact than the Newtonian reflector.

Celestial sphere. An imaginary sphere surrounding the Earth, whose centre is the same as that of the Earth's globe.

Cepheid. A short-period variable star, very regular in behaviour; the name comes from the prototype star, Delta Cephei. Cepheids are astronomically important because there is a definite law linking their variation periods with their real luminosities, so that their distances may be obtained by sheer observation.

Chromatic aberration. A defect in all lenses, due to the fact that light is a mixture of all wavelengths – and these wavelengths are refracted unequally, so that false colour is produced round a bright object such as a star. The fault may be reduced by making the lens a compound arrangement, using different kinds of glasses.

Chromosphere. That part of the Sun's atmosphere which lies above the bright surface or photosphere.

Circumpolar star. A star which never sets. For instance, Ursa Major (the Great Bear) is circumpolar as seen from England; Crux Australis (the Southern Cross) is circumpolar as seen from New Zealand.

Conjunction. (1) A planet is said to be in conjunction with a star, or with another planet, when the two bodies are apparently close together in the sky. (2) For the inferior planets, Mercury and Venus, inferior conjunction occurs when the planet is approximately between the Earth and the Sun; superior conjunction, when the planet is on the far side of the Sun and the three bodies are again lined up. Planets beyond the Earth's orbit can never come to inferior conjunction, for obvious reasons.

Corona. The outermost part of the Sun's atmosphere, made up of very tenuous gas. It is visible with the naked eye only during a total solar eclipse.

Coronagraph. A device used for studying the inner corona at times of non eclipse.

Cosmic rays. High-velocity particles reaching the Earth from outer space. The heavier cosmic-ray particles are broken up when they enter the upper atmosphere.

Cosmic year. The time taken for the Sun to complete one revolution round the centre of the Galaxy: about 225,000,000 years.

Cosmology. The study of the universe considered as a whole.

Counterglow. The English name for the sky-glow more generally called by its German name of the Gegenschein.

Day, sidereal. The interval between successive meridian passages, or culminations, of the same star: 23 h 56 m 4 s.091.

Day, solar. The mean interval between successive meridian passages of the Sun: 24 h 3 m 56 s.555 of mean sidereal time. It is longer than the sidereal day because the Sun seems to move eastward against the stars at an average rate of approximately one degree per day.

Declination. The angular distance of a celestial body north or south of the celestial equator. It corresponds to latitude on the Earth.

Dewcap. An open tube fitted to the upper end of a refracting telescope. Its role is to prevent condensation upon the object-glass.

Doppler Effect. The apparent change in wavelength of the light from a luminous body that is in motion relative to the observer. With an approaching object, the wavelength is apparently shortened, and the spectral lines are shifted to the blue end of the spectral band; with a receding body there is a red shift, since the wavelength is apparently lengthened.

Double star. A star made up of two components – either genuinely associated (binary systems) or merely lined up by chance (optical pairs).

Driving clock. A mechanism for driving a telescope round at a rate which compensates for the axial rotation of the Earth, so that the object under observation remains fixed in the field of view.

Earthshine. The faint luminosity on the night side of the Moon, frequently seen when the Moon is in its crescent phase. It is due to light reflected on to the Moon from the Earth.

Eclipse, lunar. The passage of the Moon through the shadow cast by the Earth. Lunar eclipses may be either total or partial. At some eclipses, totality may last for approximately l ¾ hours, though most are shorter.

Eclipse, solar. The blotting-out of the Sun by the Moon, so that the Moon is then directly between the Earth and the Sun. Total eclipses can last for over 7 minutes under exceptionally favourable circumstances. In a partial eclipse, the Sun is in completely covered. In an annular eclipse, exact alignment occurs when the Moon is in the far part of its orbit, and so appears smaller than the Sun; a ring of sunlight is left showing round the dark body of the Moon. Strictly speaking, a solar 'eclipse' is the occultation of the Sun by the Moon.

Eclipsing variable (or Eclipsing binary). A binary star in which one component is regularly occulted by the other, so that the total light which we receive from the system is reduced. The prototype eclipsing variable is Algol (Beta Persei).

Ecliptic. The apparent yearly path of the Sun among the stars. It is more accurately defined as the projection of the Earth's orbit on to the celestial sphere.

Elongation. The angular distance of a planet from the Sun, or of a satellite from its primary planet.

Equator, celestial. The projection of the Earth's equator on to the celestial sphere.

Equatorial mounting for a telescope. A mounting in which the telescope is set up on an axis which is parallel with the axis of the Earth. This means that one movement only (east to west) will suffice to keep an object in the field of view.

Equinox. The equinoxes are the two points at which the ecliptic cuts the celestial equator. The vernal equinox or First Point of Aries now lies in the constellation of Pisces; the Sun crosses it about 21 March each year. The autumnal equinox is known as the First Point of Libra; the Sun reaches it about ?? September yearly.

Escape velocity. The minimum velocity which an object must have in order to escape from the surface of a planet, or other celestial body, without being given any extra impetus.

Eyepiece (or Ocular). The lens, or combination of lenses, at the eye-end of a telescope. It is responsible for magnifying the image of the object under study. With a positive eyepiece (for instance, a Ramsden, Orthoscopic or Monocentric) the image plane lies between the eyepiece and the object glass (or main mirror); with a negative eyepiece (such as a Huyghenian or Tolles) the image plane lies inside the eyepiece. A Barlow lens is concave, and is mounted in a short tube which may be placed between the eyepiece and the object-glass (or mirror). It increases the effective focal length of the telescope, thereby providing increased magnification.

Faculæ. Bright, temporary patches on the surface of the sun.

Finder. A small, wide-field telescope attached to a larger one, used for sighting purposes.

Fireball. A brilliant meteor. There is no set definition, but a meteor with a magnitude of brighter than –5 will be classed as a fireball.

Flares, solar. Brilliant eruptions in the outer part of the Sun's atmosphere. Normally they can be detected only by spectroscopic means (or the equivalent), though a few have been seen in integrated light. They are made up of hydrogen, and emit charged particles which may later reach the Earth, producing magnetic storms and displays of auroræ. Flares are generally, though not always, associated with sunspot groups.

Flocculi. Patches of the Sun's surface, observable with spectroscopic equipment. They are of two main kinds: bright (calcium) and dark (hydrogen).

Galaxies. Systems made up of stars, nebulæ, and interstellar matter. Many, though by no means all, are spiral in form.

Galaxy, the. The system of which our Sun is a member. It contains approximately 100,000 million stars, and is a rather loose spiral.

Gamma-rays. Radiation of extremely short wavelength.

Gegenschein. A faint sky-glow, opposite to the Sun and very difficult to observe. It is due to thinly-spread interplanetary material.

Gibbous phase. The phase of the Moon or planet when between half and full.

Geosynchronous orbit. An orbit round the Earth at an altitude of 35 900 km, where the period will be the same as the Earth's sidereal rotation period – 23h 56m 4.1s – assuming that the orbit is circular and lies in the plane of the Earth's equator.

Globules. Small dark patches inside gaseous nebulæ. They are probably embryo stars.

Great circle. A circle on the surface of a sphere whose plane passes through the centre of that sphere.

Green Flash. Sudden, brief green light seen as the last segment of the Sun disappears below the horizon. It is purely an effect of the Earth's atmosphere. Venus has also been known to show a Green Flash.

Heliosphere. The area round the Sun extending to between 50 and 100 A.U. where the Sun's influence is dominant. The boundary, where the solar wind merges with the interstellar medium, is called the heliopause.

Herschelian reflector. An obsolete type of telescope in which the main mirror is tilted, thus removing the need for a secondary mirror.

Hertzsprung–Russell diagram (usually known as the H–R Diagram). A diagram in which stars are plotted according to the spectral types and their absolute magnitudes.

Inferior planets. Mercury and Venus, whose distances from the Sun are less than that of the Earth.

Infra-red radiation. Radiation with wavelength longer than that of visible light (approximately 7500 Ångströms).

Interferometer, stellar. An instrument for measuring star diameters. The principle is based upon light-interference.

Ionosphere. The region of the Earth's atmosphere lying above the stratosphere.

Kelvin scale. A scale of temperature. 1 K is equal to 1 °C, but the Kelvin scale starts at absolute zero (–273.16°C).

Kepler's laws of planetary motion. These were laid down by Johannes Kepler, from 1609 to 1618. They are: (1) The planets move in elliptical orbits, with the Sun occupying one focus. (2) The radius vector, or imaginary line joining the centre of the planet to the centre of the Sun, sweeps out equal areas in equal times. (3) With a planet, the square of the sidereal period is proportional to the cube of the mean distance from the Sun.

Kiloparsec. One thousand parsecs (3260 light years).

Kirkwood gaps. Gaps in the main asteroid belt, where the periods would be commensurate with that of Jupiter – so that Jupiter keeps these areas 'swept clear'.

Neutron star. The remnant of a massive star which has exploded as a supernova. Neutron stars send out rapidly-varying radio emissions, and are therefore called 'pulsars'. Only two (the Crab and Vela pulsars) have as yet been identified with optical objects.

Newtonian reflector. A reflecting telescope in which the light is collected by a main mirror, reflected on to a smaller flat mirror set at an angle of 45°, and thence to the side of the tube.

Nova. A star which suddenly flares up to many times its normal brilliancy, remaining bright for a relatively short time before fading back to obscurity.

Object-glass (or Objective). The main lens of a refracting telescope.

Occultation. The covering-up of one celestial body by another.

Ocular. Alternative name for a telescope eyepiece.

Oort cloud. An assumed spherical shell of comets surrounding the Solar system, at a range of around one light-year.

Opposition. The position of a planet when exactly opposite to the Sun in the sky; the Sun, the Earth and the planet are then approximately lined up.

Orbit. The path of a celestial object.

Parallax, trigonometrical. The apparent shift of an object when observed from two different directions.

Parsec. The distance at which a star would have a parallax of one second of arc: 3.26 light years, 206 265 astronomical units, or 30.857 million million kilometres

Penumbra. (1) The area of partial shadow to either side of the main cone of shadow cast by the Earth. (2) The lighter part of a sunspot.

Perigee. The position of the Moon in its orbit when closest to the Earth.

Perihelion. The position in orbit of a planet or other body when closest to the Sun.

Phases. The apparent changes in shape of the Moon and the inferior planets from new to full. Mars may show a gibbous phase, but with the other planets there are no appreciable phases as seen from Earth.

Photometer. An instrument used to measure the intensity of light from any particular source.

Photometry. The measurement of the intensity of light.

Photosphere. The bright surface of the Sun.

Planetary nebula. A small, dense, hot star surrounded by a shell of gas. The name is ill-chosen, since planetary nebulæ are neither planets nor nebulæ!

Poles, celestial. The north and south points of the celestial sphere.

Position angle. The apparent direction of one object with reference to another, measured from the north point of the main object through east, south and west.

Precession. The apparent slow movement of the celestial poles. This also means a shift of the celestial equator, and hence of the equinoxes; the vernal equinox moves by 50 sec of arc yearly, and has moved out of Aries into Pisces. Precession is due to the pull of the Moon and Sun on the Earth's equatorial bulge.

Prime Meridian. The meridian on the Earth's surface which passes through the Airy Transit Circle at Greenwich Observatory. It is taken as longitude 0°.

Prominences. Masses of glowing gas rising from the surface of the Sun. They are made up chiefly of hydrogen.

Proper motion, stellar. The individual movement of a star on the celestial sphere.

Protoplanet. A body forming by the accretion of material, which will ultimately develop into a planet.

Pulsar. A rotating neutron star, often a strong radio source. Not all pulsars can be detected by radio, since the radiation is emitted in beams, and it depends upon whether these beams sweep over the Earth.

Quadrant. An ancient astronomical instrument used for measuring the apparent positions of celestial bodies.

Quasar. The core of a very powerful, remote active galaxy. The term QSO (quasi-stellar object) is also used.

Radial velocity. The movement of a celestial body toward or away from the observer; positive if receding, negative if approaching.

Radiant. The point in the sky from which the meteors of any particular shower seem to radiate.

Retrograde motion. Orbital or rotational movement in the sense opposite to that of the Earth's motion.

Reversing layer. The gaseous layer above the Sun's photosphere.

Right ascension. The angular distance of a celestial body from the vernal equinox, measured eastward. It is usually given in hours, minutes and seconds of time, so that the right ascension is the time-difference between the culmination of the vernal equinox and the culmination of the body.

Schmidt camera (or Schmidt telescope). An instrument which collects its light by means of a spherical mirror; a correcting plate is placed at the top of the tube. It is a purely photographic instrument.

Scintillation. Twinkling of a star; it is due to the Earth's atmosphere. Planets may also show scintillation when low in the sky.

Sextant. An instrument used for measuring the altitude of a celestial object.

Sidereal period. The revolution period of a planet round the Sun, or of a satellite round its primary planet.

Sidereal time. The local time reckoned according to the apparent rotation of the celestial sphere. When the vernal equinox crosses the observer's meridian, the sidereal time is 0 hours.

Solar nebula. The cloud of interstellar gas and dust from which the Solar System was formed – around 5000 million years ago.

Solar wind. A flow of atomic particles streaming out constantly from the Sun in all directions.

Solstices. The times when the Sun is at its maximum declination of approximately 23½ degrees; around 22 June (summer solstice, with the Sun in the northern hemisphere of the sky) and 22 December (winter solstice, Sun in the southern hemisphere).

Spectroscopic binary. A binary system whose components are too close together to be seen individually, but which can be studied by means of spectroscopic analysis.

Speculum. The main mirror of a reflecting telescope.

Spherical aberration. Blurring of a telescope image; it is due to the fact that the lens (or mirror) does not bring the light-rays falling on its edge and on its centre to exactly the same focal point.

Superior planets. All the planets lying beyond the orbit of the Earth in the Solar System (that is to say, all the principal planets apart from Mercury and Venus).

Supernova. A colossal stellar outburst, involving (1) the total destruction of the white dwarf member of a binary system, or (2) the collapse of a very massive star.

Tektites. Small, glassy objects found in a few localized areas of the Earth. They are not now believed to be meteoritic.

Terminator. The boundary between the day- and night-hemispheres of the Moon or a planet.

Transit. (1) The passage of a celestial body across the observer's meridian. (2) The projection of Mercury or Venus against the face of the Sun.

Umbra. (1) The main cone of shadow cast by the Earth. (2) The darkest part of a sunspot.

Variable stars. Stars which change in brilliancy over short periods. They are of various types.

White dwarf. A very small, very dense star which has used up its nuclear energy, and is in a very late stage of its evolution.

Year. (1) **Sidereal:** the period taken for the Earth to complete one journey round the Sun (365.26 days). (2) **Tropical:** the interval between successive passages of the Sun across the vernal equinox (365.24 days). (3) **Anomalistic:** the interval between successive perihelion passages of the Earth (365.26 days; slightly less than 5 minutes longer than the sidereal year, because the position of the perihelion point moves along the Earth's orbit by about 11 seconds of are every year). (4) **Calendar:** the mean length of the year according to the Gregorian calendar (365.24 days, or 365 d 5 h 49 m 12 s).

Zenith. The observer's overhead point (altitude 90 degrees).

Zodiac. A belt stretching round the sky, 8 degrees to either side of the ecliptic, in which the Sun, Moon and principal planets are to be found at any time. (Pluto and many asteroids can leave the Zodiac.)

Zodiacal Light. A cone of light rising from the horizon and stretching along the ecliptic; visible only when the Sun is a little way below the horizon. It is due to thinly spread interplanetary material near the main plane of the Solar System.

INDEX